House Spiders - Worldwide

Wolfgang Nentwig · Jutta Ansorg ·
Paula E. Cushing ·
Yvonne Kranz-Baltensperger ·
Christian Kropf

House Spiders - Worldwide

 Springer

Wolfgang Nentwig
Institute of Ecology and Evolution
University of Bern
Bern, Switzerland

Paula E. Cushing
Denver Museum of Nature and Science
Denver, CO, USA

Christian Kropf
Natural History Museum Basel
Basel, Switzerland

Jutta Ansorg
Freienwil, Aargau, Switzerland

Yvonne Kranz-Baltensperger
Natural History Museum Bern
Bern, Switzerland

ISBN 978-3-031-70447-5   ISBN 978-3-031-70448-2   (eBook)
https://doi.org/10.1007/978-3-031-70448-2

The original submitted manuscript has been translated into English. The translation was done using artificial intelligence (machine translation by the service DeepL.com). A subsequent revision was performed by the authors in collaboration with Dr. Jason Dunlop (Museum für Naturkunde, Leibnitz Institute for Evolution and Biodiversitry Science, Berlin, Germany) in order to refine the work linguistically and technically.

Translation from the German language edition: "Hausspinnen weltweit" by Wolfgang Nentwig et al., © Förderverein für Spinnenforschung 2024. Published by Springer Berlin Heidelberg. All Rights Reserved.

The cover illustration shows a female of the common spitting spider *Scytodes thoracica*. Photo by Jean-Philippe Taberlet.

This Springer imprint is published by the registered company Springer Nature Switzerland AG
The registered company address is: Gewerbestrasse 11, 6330 Cham, Switzerland

If disposing of this product, please recycle the paper.

Preface

First a clarification: house spiders are not pets. They are those spider species that live with us in our apartments and houses. They are essentially our subtenants and not just passing through (although there are also such temporary visitors). At first glance, this is not surprising, because spiders live everywhere on earth, so why not in our homes as well, especially since some spider species are indeed not too demanding in terms of their habitat.

And yet it is surprising that many spider species apparently feel comfortable in our dwellings. Surprising because the microclimate in our buildings in many parts of the world is simply not suitable for spiders. Humidity in our buildings is much too low, often it is too clean (yes, that's possible) and the food supply is too irregular, not to mention the dangers of human house sweeping and vacuuming. To be able to cope with this, one must have quite a few survival tricks up one's sleeve.

Since human buildings across the world have certain similarities, the motto is that whoever can survive in a house in one place can probably also cope in our homes in another region. The spiders just need to be able to reach this next place. However, this problem is actually negligibly small, because humans have been helping their house spiders for centuries, enabling them to reach other buildings, settlement areas, countries, and even continents. We do not consciously transport spiders overseas, but when spider webs with animals or egg cocoons stick onto and into the vessels, sacks, wooden crates, and containers that we move around the world, who even notices?

Our house spiders could not only tell great stories of surviving in a difficult environment, they also know human conditions of import and export, at least from the perspective of stowaways. This only became significant about 500 years ago, at least as far as overseas traffic is concerned, so the history of our spreading house spiders also somewhat reflects the triumph of sailing ships, which began with Columbus. This seafaring is closely linked to European colonial history. Therefore, today we find the most common European house spiders on many other continents, but African, American, and Asian spider species have also made it to Europe and have indeed achieved a worldwide distribution. Who thinks of such a thing when it comes to simple house spiders?

Some people are fascinated by their eight-legged house guests. Others want to ensure that no danger emanates from them. And that's why we, as members of the Association for the Promotion of Spider Research, like most spider researchers,

keep getting identification inquiries about house spiders. After our first, very successful book *All You Need to Know About Spiders*, which gives a general overview
of the lifestyle of spiders, we would like to introduce, in this new book, the most
common species of house spiders found throughout the world and to paint a portrait
of the most common species you are likely to encounter in your own homes.

We wish you much pleasure in reading this book and hope that through it you
will discover the joy in welcoming these eight-legged guests into your own home or
will, at least, consider not evicting them immediately.

Association for the Promotion of Spider Research
Bern, Switzerland   Wolfgang Nentwig
Freienwil, Switzerland   Jutta Ansorg
Denver, CO, USA   Paula E. Cushing
Bern, Switzerland   Yvonne Kranz-Baltensperger
Basel, Switzerland   Christian Kropf

The Association for the Promotion of Spider Research

The Association for Spider Research, founded in Switzerland in 2016, aims to inform the public about the importance of spiders and promote scientific research. The Association works closely with natural history museums and arachnological professional associations. Currently, three projects are at the forefront of our work: (1) Support of the online identification page Spiders of Europe (https://araneae.nmbe.ch), which offers identification keys, features, illustrations, and distribution maps for around 5000 European spider species in a German and an English version; (2) Support of the World Spider Catalog (https://wsc.nmbe.ch), which contains all taxonomically relevant information on the over 52,000 spider species occurring worldwide as well as the associated professional information (around 17,500 professional articles); (3) Knowledge transfer in book form, like this book here, which is also available in German, to inform as many interested people as possible about spiders in an approachable way. The proceeds of this book go entirely to the projects of the association.

For more information, please visit https://wsc.nmbe.ch/association/index. If you would like to support the Association for the Promotion of Spider Research, please contact us.

Wolfgang Nentwig (President) (w.nentwig@gmail.com)
Ambros Hänggi (Vice President) (ambros.haenggi@sunrise.ch)

Acknowledgments

Editing of the photos was done by Alice Nentwig, and we thank her very much. Miriam Frutiger drew the eye illustrations in Chap. 4. Ambros Hänggi has supported our project in various ways and undertook, together with Anne-Sarah Ganske, Anna Stäubli, and Karin Urfer, proofreading of the manuscript. We also extend our heartfelt thanks to them. Special thanks go to Jason Dunlop for checking the English translation.

Many colleagues have informed us about which spider species live in their homes. For this, we thank Anita Aisenberg, Igor Armiach, Yuki Baba, Francesco Ballarin, Nicky Bay, Robb Bennett, Rich Bradley, Antonio Brescovit, John Caleb, Pedro Cardoso, Reginald Christiaan, Szinetar Csaba, Tarik Danisman, Christo Deltshev, Ansie Dippenaar-Schoeman, Petr Dolejs, Marc Domènech, Anka Eichhoff, Hisham El-Hennawy, Sergei Esyunin, Arne Fjellberg, Volker Framenau, Holger Frick, Fulvio Gasparo, Efrat Gavish-Regev, Marcelo Gonzaga, Gordana Grbic, Matt Greenstone, Cristian Grismado, Charles Haddad, Vic Hamilton-Attwell, Marc Harvey, Marie Herberstein, Linda Hernandez Duran, Marco Isaia, Peter Jäger, Allen Jones, Anike Kirsten, Joseph K. H. Koh, Marjan Komnenov, Gábor Kovács, Torbjörn Kronestedt, Kadir Bogac Kunt, Matjaz Kuntner, Daiquin Li, Richard Louvigny, Nuria Macias, Ivan Magalhaes, Jagoba Malumbres-Olarte, Stefano Mammola, Yura Marusik, Jim McLean, Kirill Mikhailov, Majid Moradmand, Ed Nieuwenhuys, Myles Nolan, Hirotsugu Ono, Paolo Pantini, Cristina Reims, Paul Selden, Stano Pekar, Pham Dinh Sac, Luis Piacentini, Moothedathu S. Pradeep, Martin Ramirez, Adalberto Santos, Anna Sestakova, Osman Seyyar, Petra Sierwald, Anna Stäubli, Puthoor Pattammal Sudhin, Andres Taucare Rios, Debbie Taylor, Alessio Trotta, Wynand Uys, Rick West, Jonathan Whitaker, Jörg Wunderlich, Xin Xu, Alireza Zamani, and Christos Zoumides, as well as the Ecdysis portal for live-data arthropod collections (https://ecdysis.org/) and the Denver Museum of Nature & Science arachnology collection.

For the provision of photographic material, we thank Gordon Ackermann, Ewa Ansorg, Gilles Arbour, Kyron Basu, Mark A. Brogie, Reinhard Bülte, Ken Childs, Cesar Crash, Mike Deep, Eckhart Derschmidt, Benjamin Eggs, Volker Framenau, Sarah Friedli, John P. Friel, Guido Gabriel, Ambros Hänggi, Arnaud Henrard, Michael Hohner, Bernhard Huber, Jim T. Johnson, Barbara Knoflach-Thaler, Joseph K. H. Koh, Hans-Ulrich Kohler, Gábor Kovács, Matjaz Kuntner, Alvora Laborda, Pierre Loria, John Maxwell, Ondřej Michálek, Pierre Oger, Jörg Pageler, Walter

Pfliegler, Bastian Rast, Sarah Rose, Dragiša Savić, Michael Schäfer, Miguel Simo, Jean-Philippe Taberlet, Vida van der Walt, Rick Vetter, Norman Wilson, Hartmut Wisch, Brandon Woo, and Samuel Zschokke.

We also thank Stefanie Wolf, Kerstin Barton, and Lars Koerner from Springer Publishing for their cooperation, interest, and support of this project.

Disclaimer

The borders shown on the distribution maps are for orientation purposes only and do not always represent national borders. In addition, not all current official or de facto national borders are shown consistently. No political message is associated with the distribution maps.

Contents

About the Authors

Jutta Ansorg (jutta.ansorg@yahoo.com) studied Energy and Process Engineering and received her doctorate in 2001 from the Technical University of Berlin. From 1993 to 1998, she was a Lecturer for Practical Mathematics and conducted research in the field of process engineering. From 1998 to 2014, she worked as an engineer in Switzerland for international power plant and waste incineration plant construction. Since 2014, she has been a specialist for air pollution control and noise protection in the Canton of Aargau (Switzerland). She has been studying spiders in her spare time for about 45 years.

Paula E. Cushing (paula.cushing@dmns.org) earned her Doctorate in Zoology from the University of Florida in 1995. She has been researching spiders and publishing about spiders, solifuges, and scorpions for 40 years. Since 1998, she has been the curator for Invertebrate Zoology at the Denver Museum of Nature & Science in the United States. She is very active in the American Arachnological Society and the International Society of Arachnology and has served as President of both associations in the past. She has also hosted several arachnological conferences and mentored young arachnologists.

Yvonne Kranz-Baltensperger (yvonne.kranz@nmbe.ch) studied Biology at the University of Bern, completing her dissertation in 2014. Since 1997, she has been a scientific assistant for the collections in the departments of Arachnology and Entomology at the Natural History Museum Bern. She has conducted research as part of the PBI project (Planetary Biodiversity Inventory) on goblin spiders (Oonopidae).

Christian Kropf (christian.kropf@bs.ch) studied Biology and received his PhD 1992 at the University of Graz (Austria). He was Assistant at the University of Graz 1992–95 and Curator for invertebrate animals at the Natural History Museum Bern (Switzerland) 1995–2023. Since 2024 he is Curator of Biological Sciences at the Natural History Museum Basel (Switzerland). Since 1996 he is Lecturer at the University of Bern where he also received his Habilitation 2011. Research focuses on taxonomy, systematics, and functional morphology of arthropods, especially spiders. He is co-editor of "araneae - Spiders of Europe" and co-organizer of the World Spider Catalog.

Wolfgang Nentwig (w.nentwig@gmail.com) studied Biology and received his PhD in 1981 at the University of Marburg (Germany). He was visiting researcher at the Smithsonian Tropical Research Institute (Panama 1983/1984) and 1985–1988 Assistant Professor at the University of Regensburg (Germany). 1988–2019 he was Professor of Ecology, University of Bern (Switzerland). His research focused on agroecology, invasion ecology, spider ecology, and spider venom. co-founder of "araneae—Spiders of Europe" and co-organizer of the World Spider Catalog.

Part I

General Part

Understanding and Recognizing Spiders

1

When we, as spider researchers, talk to laypeople about our favorite animals, two completely different reactions are normal: (1) Ah, these are the animals that build those beautiful webs (they always mean orb webs); or (2) Oh my goodness, these are the big, ugly, hairy creatures that lurk in our bathrooms and are even venomous. We, therefore, regularly experience the fact that we are sure to attract attention, as the fascination with spiders is always there, whether it is positive or negative. But what exactly do spiders look like, and what is so special about them? In this introductory chapter, we describe the most important features of the structure and life of spiders. For those who want to know more about this fascinating group of animals, we are happy to refer you to our introductory book, "All You Need to Know About Spiders."

1.1   Eight Long Legs and Two Body Halves

Spiders are easy to recognize. They have two body parts, which are connected by an extremely slim waist. On the **front body** (prosoma), we find four multi-segmented pairs of legs, totaling eight legs, and two forward-facing **pedipalps** that look like shortened legs (Figs. 1.1 and 1.2). In addition, there are usually two downward-facing **chelicerae**, also called the jaws, with a movable **fang** (Fig. 1.3). On the front body, we often see eight **eyes** at the front, but sometimes only six or fewer. On its underside, there is a solid plate (sternum).

The **hind body** (opisthosoma) can have beautiful markings on the back, such as an elongated to triangular spot in the front area, which is referred to as a **heart spot**. On the underside, you can see the lung covers and, in females, usually structures around the genital opening (epigyne). At the end of the hind body are the **spinnerets**, from which the spider's silk emerges. The combination of the front and hind body, the arrangement of the legs at the front, and the spinnerets at the back allow spiders a unique degree of mobility in web building and prey capture, or during mating.

© Association for the Promotion of Spider Research 2024
W. Nentwig et al., *House Spiders - Worldwide*,
https://doi.org/10.1007/978-3-031-70448-2_1

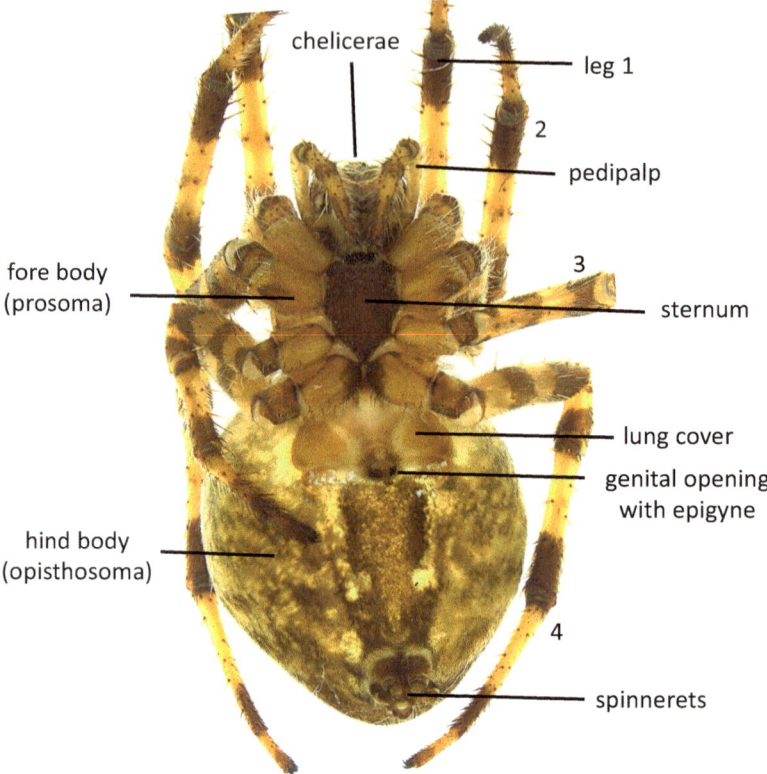

Fig. 1.1 Ventral view of a female European garden spider *Araneus diadematus* (orb-weavers, Araneidae). (*Photo* Yvonne Kranz-Baltensperger)

Based on these few characteristics, spiders can also be unmistakably recognized as such. However, caution is advised with most spiders, as they are often very small and can be easily damaged. Therefore, a close look or a good magnifying glass, and sometimes even a microscope, is needed. With the exception of some large species, including the well-known tarantulas, spiders have a **body length** of at most 40 mm (1½ in.). This is measured from the front edge of the front body to the rear edge of the hind body, excluding the legs. The leg span can be significantly larger, but with our house spiders, it is hardly more than about 10 cm. The huge, thick, black, hairy spider that hangs on the wall at night or sits in the bathtub is always a product of our imagination.

Two more groups should be excluded: harvestmen and mites are not spiders, as they only have one body part without the thin "wasp waist." Both are, however, grouped with spiders, scorpions, and some other animals to form the arachnids. Finally, a look at insects: unlike spiders, their body is divided into a head, thorax, and abdomen; they have only six legs; they have a pair of antennae on the head; and they usually have two compound eyes.

Fig. 1.2 Dorsal view of a male European garden spider *Araneus diadematus* (orb-weavers, Araneidae). (*Photo* Yvonne Kranz-Baltensperger)

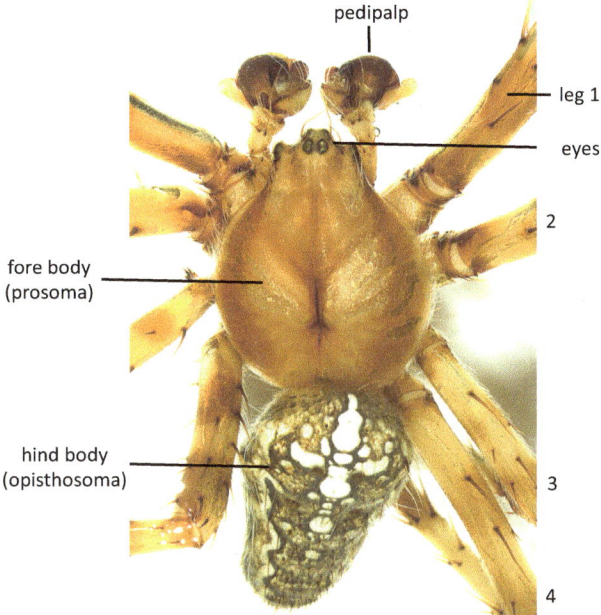

pedipalp

leg 1

eyes

2

fore body (prosoma)

hind body (opisthosoma)

3

4

Fig. 1.3 Front view of a male European garden spider *Araneus diadematus* (orb-weavers, Araneidae). (*Photo* Yvonne Kranz-Baltensperger)

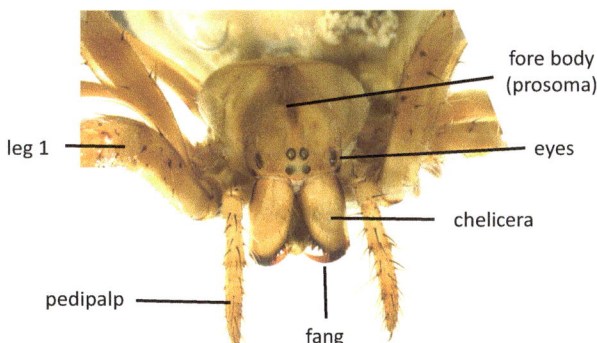

fore body (prosoma)

leg 1

eyes

chelicera

pedipalp

fang

1.2   Legs: Not Just for Walking

The long legs of spiders are composed of seven movable joints connected together. This gives us a hint as to where in the animal kingdom spiders can be classified: among the arthropods, which also include insects, crustaceans, and millipedes. All these animals have an **exoskeleton**; in other words, their body shape is determined by the rigid outer shell, the **cuticle**.

In spiders, the front body and legs are hard, while the hind body is covered by a thinner shell or exoskeleton. As a result, if the spider wants to grow, it must shed the rigid outer exoskeleton. To do this, a new, still soft outer covering is formed under the old shell. During the **molt**, only the cuticle on the front body bursts open. The body and legs are then pulled out of the old, firm shell. Since spiders now pump some of their body fluid (hemolymph) from the hind body into the front body, the new, still soft cuticle can be stretched out to a larger size than it was before. In the larger form, the cuticle hardens, so that the spider grows with each molt. For the spider, this is an extremely difficult moment, because until the new exoskeleton has hardened over the course of 1 to 2 days, the spider is usually defenseless against its enemies.

Spider legs bear many sensory organs and various **hairs**, **bristles**, or **spines**. Spider hairs are different from our hairs and are called setae. Even though spiders (usually) have eight eyes, they typically orient themselves in their environment through the sensory organs on their legs. They mainly perceive vibrations from the ground, from the spider web, and from the air, and in this way, they know whether a possible prey, an enemy, or a sexual partner is approaching. With specialized hairs, they can also measure temperature and humidity, as well as smell or taste chemical substances.

To understand the way spiders move, one must look at their legs more closely. All spiders have **claws** at the tip of the foot, either two claws (free-hunting species) or three (usually web-building species). The third claw of the web-building species is necessary so that the spider can hold onto its web threads. For both groups, the following rule applies: with the fine claws, they can cling to the smallest structures and easily climb up tree bark, as well as house walls.

However, on very smooth surfaces like glass panes or bathtubs, this does not work. The large spider that we occasionally find helpless in the bath is a web-building species. With its claws, it can move optimally in its web, but they are not useful for the smooth bathtub wall, and the spider just keeps slipping down. Other spiders, such as jumping spiders, have no problem with smooth walls. They have tufts of hundreds of tiny, slightly flattened **adhesive hairs** (scopulae), which allow them to move on shiny surfaces like mirrors and even to hang upside down on the ceiling.

1.3   Spider Silk: A High-Tech Product for All Situations

Spider silk is extremely diverse, as spiders have several types of glands in their hind body that can produce secretions with different properties. Only when the liquid substances emerge from the various **spinnerets** do they solidify and become silk threads with different physical properties under tension. Spider silk is used by the mother to protect the eggs in a cocoon, spiders build different types of catching devices (i.e., spider webs), residential webs that serve as a retreat, and wandering spiders always produce a safety line that allows them to crawl back up in the event of a possible fall.

Yes, small spiders can even fly thanks to their spider silk, because during favorable weather, the hind body is raised up into the air and fine silk threads emerge from the spinnerets. These are captured by rising air, and when the threads are long enough, the small spiders are blown away (**ballooning**). In the Earth's electrostatic field, the negatively charged spider silk is also repelled by the Earth. This is an important strategy for spiders to disperse into new areas, but it has a significant disadvantage: the spider cannot determine the landing point itself.

The most primitive **webs** are simple tripwires that pick up the movement of passing prey and relay it to the **living tube** where the spider lurks. We find such webs among the house spiders discussed here, in, for example, the case of the tube-dwelling spiders (Segestriidae). A greater effort in silk production, and thus more compact webs, can be found in the funnel-web spiders (Agelenidae) or the sheet weavers (Linyphiidae). Particularly sophisticated are catching techniques in which individual silk threads are equipped with adhesive droplets. We find these on the spiral thread of the orb web in the orb-weavers (Araneidae) and on individual threads at the margins of the web in the cobweb spiders (Theridiidae).

The hackled mesh-weavers (Amaurobiidae), for example, catch their prey with another trick. They pull ultra-thin silk threads (the cribellum wool or crimped threads) from spinnerets transformed into a plate shape (the cribellum) onto normal silk threads with a comb-like row of particularly strong bristles on the fourth pair of legs (called the calamistrum). On the one hand, these fine threads work in the same way that cotton wool sticks to our fingertips. On the other hand, these strongly crimped, so-called **cribellate** threads absorb oily-fatty, i.e., soluble, components from the wax coating of the cuticle of the prey, such that an insoluble connection is created between the prey's own body and the web. Cribellate-catching threads are therefore considered just as efficient as adhesive droplets (used, for example, by the orb-weavers, Araneidae). For example, we also find such cribellate webs for prey capture in mesh-weavers (Dictynidae), crevice weavers (Filistatidae), or hackled orb-weavers (Uloboridae).

1.4   Spiders Do Not Eat, They Drink

Once a spider has caught its prey, it is often wrapped and bound in silk. This happens with many web-building spiders. Free-roaming spiders often seize the prey with several of their legs and hold it tight. To paralyze the prey, a venomous bite is then applied. For this, spiders have paired **venom glands** in the front body. The venom is channeled through the basal part of the jaws to the tips of the fangs and injected into the prey. So, the vast majority of spiders are venomous!

However, **spider venom** consists of a mixture of many individual components and evolved in spiders specifically to work against insects and other invertebrates. Spiders also produce only very small amounts of their venom. Therefore, spider venom usually poses no danger to humans, as it is usually only present in insufficient quantities. Furthermore, spiders are not aggressive toward humans, and most species cannot penetrate human skin with their small fangs.

When the prey item is immobile or dead, it is not eaten in the way one would expect from, for example, insects. Insects have complex mouthparts that allow them to dismember and swallow their prey piece by piece. By contrast, spiders have specialized in digesting their prey before the tissue enters the mouth opening. To do this, they regurgitate **digestive juice**, which dissolves the internal organs of the prey. The tissue of the prey item is thus completely liquefied and can then be more or less completely sucked up by the spider. The actual digestion process, therefore, takes place outside the spider body. This is also referred to as **extraintestinal digestion**.

In this way, spiders not only consume prey but also old parts of their webs that have become dusty or destroyed. This spider silk is digested and immediately reused for the production of new silk. Experiments with radioactively labeled spider silk have shown that the same building blocks from the digested silk reappear in the newly produced spider threads and that this recycling can happen very quickly.

1.5   Eating and Being Eaten

Detailed information about the hunting behavior of spiders can be found in the chapters with the species descriptions. In principle, there are two strategies used by spiders: setting traps, i.e., building webs for prey capture, or free hunting. Among the trap setters, different types of webs occur. Orb webs are probably well known to everyone. They can be very small, but sometimes they have frame threads several meters long, especially near buildings. Sheet webs, in all their variations, can be found in basements where the giant house spider *Eratigena atrica* lives. Small sheet webs can be found associated with the wall spider *Brigittea civica*.

Other spiders catch prey as hunters, albeit without tools but with a lot of physical strength. Mention should be made here of jumping spiders, which have correspondingly well-developed, powerful eyes. With these, they can recognize prey at a certain distance and also estimate the distance to the prey. This is important because the spider jumps at the prey with a leap of up to 10 times its body length and then immediately applies a venomous bite. The jump can sometimes fail, and the spider may fall. But that's not a problem, because the spider always produces a safety thread that it can climb back up on. An interesting fact: the production of the safety thread runs as fast as the speed of the jump!

Spiders, as effective predators, play an important role in the balance of nature. They devour a large number of prey animals, among which are important plant pests, mosquitoes, or other nuisances. However, spiders are hardly food specialists and therefore do not distinguish between beneficial organisms and pests (a pair of terms that only makes sense from a human perspective), but also eat other predators and, of course, spiders. Therefore, they are not able to eradicate a given pest in a certain region, but they can reduce it to a lower and thus (for humans) less harmful density.

1.6   Sexual Behavior: Spiders Like It Complicated

Spiders are predators, and as predators, they devour everything they can overpower. They are by no means picky and do not spare their own kind. For a successful mating, it is therefore important that the partners signal to each other that they are not prey and are also willing to mate. Over the course of their evolution, spiders have developed a number of mechanisms that allow safe mating. Spiders are characterized by **complicated behavioral patterns**, but this also ensures that male spiders survive the mating process. The common image of the man-eating female spider is therefore mostly incorrect.

For the transfer of sperm from the male to the female, spiders have evolved an indirect pathway via **secondary sexual organs**. The actual male sexual organs, the testes, are located inside the hind body, while the special organs for sperm transfer are associated with the pedipalps in front of the body. For **sperm transfer**, a drop of sperm must first be secreted onto a small sperm web specially built for this purpose. In a second step, the sperm must now be taken up again by the male. For this, he uses his pedipalps, which, similar to a syringe or suction pump, take up the sperm fluid. These sperm transfer organs are usually very complicated in design and different for each species. Therefore, they help with accurate species identification.

Only after sperm uptake does the male spider approach the female. So that the male is not misunderstood as potential food, he must signal to the female that he is not prey. Again, each species has its own behaviors. In web-building spiders, the male announces himself to the female, who is in the web, by rhythmically plucking at the edge of her web. The pattern underlying the plucking is a little different for each species. In this way, the female can recognize that he is a male of her own species; otherwise, he would just be another prey item. Other species, which live on the ground, announce themselves with species-specific knocking signals with the hind body or the pedipalps on the ground. Jumping spiders, with their excellent vision, perform proper mating dances. In this case, the male dances in front of the female, raises and lowers his body, and waves with the conspicuously colored pedipalps until the female indicates she is willing to mate.

When both partners agree, the male approaches the female. For different spider families, certain procedures have been developed for this. Males can slide from the front underneath the female, while others approach from above or from the side. The male always grips the female with one or both pedipalps in order to push the tip of the pedipalp, which now contains his sperm, into or onto the female genital opening. This is located on the underside of the hind body near the front. There are also secondary sexual organs (a mating plate, also **called epigyne**, which is important for the identification of the females) that help the male to orient himself and to hook his pedipalp into the genital opening for a certain time. After a successful mating, both animals usually separate peacefully.

Apartments and houses have special climatic conditions, which allow a small group of spider species to live in our buildings. Through global goods transport, we have spread many of our house spiders throughout the world, so we often encounter the same species everywhere. There is no reason to be afraid of them, yet we describe how to remove spiders alive from the house. In general, spiders in the house are useful to humans and can be safely left there.

2.1   Climatic Conditions in a Building

Many people assume that apartments and buildings are clean and almost sterile, or at least should be, since many perceive their homes as isolated from the outside world. Of course, this is not true. The buildings in which we stay are **not germ-free** but are in **regular exchange** with their environment. So, it's no surprise that we keep finding insects in our home that have flown or run in from outside. Spiders, on the other hand, are less noticeable, even though they have already arrived in our apartments. We experts assume that in every building at least two or three spider species, each with several individuals, occur. That we don't see them at first glance is another story.

But what exactly are house spiders? In a first approximation, we can define them as those spider species that live permanently in our houses and apartments. More precisely, we must add that these are the spider species that can live in our houses. Only with certain **adaptations** is it possible for them to survive in the special conditions of a human building over the long term. House spiders are therefore essentially different species from those that live in the natural environs of a house, which we will refer to as garden spiders for the sake of simplicity. From this formulation, three important questions already arise:

- Can spiders from our garden also live in our house?
- Can house spiders survive outside?
- Why are buildings a special habitat for spiders?

The central point for understanding our house spiders is the climate in our buildings, which, at least in temperate latitudes, differs fundamentally from the climate outside with respect to frost and **dryness**. In these buildings, roofs, insulation, and heating ensure a balanced and rather dry microclimate, which is also characterized by **frost-free conditions**. Most house spiders cannot tolerate frost and would die if they had to live outdoors. Garden spiders, although they do not need frost for their well-being, have adapted to it through physiological adjustments and can survive it, whether as young or adult animals or during the egg stage (e.g., in a eggsac). Moreover, they often seek sheltered places in winter (in crevices, in leaf litter, or under stones or bark), so they are additionally protected from weather extremes.

But what prevents our garden spiders from spending the night in frost-free houses? It is the very low **humidity**, which affects spiders from outside so much that they would dry out within a few days in the building. Since cold air can absorb less water than warm air, the relative humidity in the temperate zone outside in winter is higher (on average about 75–95%) than the heated air in the house (about 40–60%). Therefore, there are two different factors that are important: garden spiders cannot live in buildings (due to dryness), and most of our house spiders cannot survive outside (due to frost).

House spiders are thus species that either originate from particularly dry habitats and are adapted to water scarcity or occur in different habitats and, as generalists, can tolerate various conditions, including our indoor climate. Specifically, the original habitats of house spiders often include desert- and semi-desert-like habitats, but also caves, rock crevices, or cavities in wood and under bark. Such habitats also exist in the **temperate zone**; therefore, house spiders originating from these regions, in addition to their pronounced resistance to dryness, can sometimes also have a certain degree of frost resistance.

And in the **tropics**? Here, frost is not an issue, but the dryness in buildings and outside can vary greatly. Native tropical spiders always live in a humid climate, and buildings are usually too dry for them. If these are then maintained and (highly) cooled with air conditioning, they quickly become uninhabitable for the local spider species. Spider species adapted to the dry indoor climate, the typical house spiders, are therefore also at an advantage here and can colonize buildings worldwide (Box 2.1).

Box 2.1: House Spiders: A Bit of Technical Jargon

Spiders that live (or can live) in areas of human settlement, and more precisely in buildings, are referred to as **synanthropic** (Greek for "with humans"). We understand **eusynanthropic** species to be those that occur exclusively in buildings and can survive there over the long term. **Hemisynanthropic** species are those that can live both in buildings and outside human settlements. Finally, **xenanthropic** species are those that occasionally come into buildings but cannot survive there. We also refer to them as incidental findings. When using these terms, the geographical context must not be forgotten, as a spider species may be able to live adequately without human settlements in a warm climate, while it can survive in very cold regions only inside buildings. In a global context, as we present it in this book, these somewhat technical terms can be confusing, and we will not use them further.

2.2   Where Do House Spiders Come From?

It's crazy, but most house spiders do not necessarily come from the immediate vicinity of our house. In Europe north of the Alps, many of today's house spiders come from the Mediterranean region and from all other continents. Thus, we encounter spiders from the Americas, Africa, and Asia within our own four walls. The same applies to other continents, where we can encounter a spider society in buildings with species originating from all over the world (Table 2.1).

Since our buildings represent a unique, highly usable, and frequently available resource for appropriately adapted spiders, these structures are populated by such spiders throughout the world. In addition to the few native species that can

Table 2.1 The 10 most common spider species found worldwide in buildings, ranked by frequency

Species	English name	Family	Originates from this continent (•) and occurs in these continents (x) in houses				
			The Americas	Europe	Africa	Asia	Australia
Pholcus phalangioides	Long-bodied cellar spider	Pholcidae	x	x	x	•	x
Parasteatoda tepidariorum	Common house spider	Theridiidae	x	x	x	•	x
Scytodes thoracica	Common spitting spider	Scytodidae	x	•	x	x	x
Steatoda triangulosa	Checkered cobweb-weaver	Theridiidae	x	•	x	x	x
Tegenaria domestica	Barn funnel-weaver	Agelenidae	x	•	x	x	x
Steatoda grossa	large false black widow	Theridiidae	x	•	x	x	x
Oecobius navus	Disc web spider	Oecobiidae	x	•	x	x	x
Hasarius adansoni	Adanson's house jumper	Salticidae	x	x	•	x	x
Eratigena atrica	Giant house spider	Agelenidae	x	•		x	
Salticus scenicus	Zebra jumping spider	Salticidae	x	•	x	x	

The most common families are discussed in Box 4.2

live in houses, the spider communities in our buildings consist of a larger number of **non-native species**, which are given access to our homes through trade and migration (see below), thereby opening up a new habitat for them. Many house spiders, therefore, de facto reach regions where they are not native through human activity. House spiders are spread by humans over a much larger area, and in extreme cases worldwide, than would be possible for the spiders on their own. The phenomenon of house spiders is thus created by humans and historically began with the settlement of new areas, construction of human habitations, and the transport of goods.

For this reason, the composition of house spiders amongst each other is very similar worldwide. However, if you compare the house spiders in some buildings with the spiders of the surrounding habitats, the spectrum of species outdoors is much larger. This means that with knowledge of the house spiders from Europe, we will find old acquaintances (at least among the spiders) in buildings everywhere on a trip around the world. This similarity of the worldwide spider societies in buildings also implies that it is quite possible to present a global view in a book about house spiders.

2.3   How Do House Spiders Get into Our Houses?

If we look at the settlement history of humans over the last 500 years, we have a more or less unconscious, limited perspective toward most cultures. Often, we see the discovery of the world and its subsequent colonization only from a European perspective. This is certainly a much too simplistic view of things in general, but it helps us in analyzing the triumphal march of our house spiders around the world.

When Europeans reached the Americas and circumnavigated the world, a **worldwide network** of trade relations soon developed between the European colonies across all the continents and Europe. With sailing ships, large quantities of goods were transported. What was transported, however, lay first in sheds and warehouses on a distant continent, was then brought onto a ship, and eventually reached warehouses again at its destination. The trade goods were well packed, but the packaging itself was often not "clean." Dirt, grime, and plant seeds clung to the outside, and cracks and crevices offered numerous opportunities for insects and spiders to hide in them.

As **stowaways**, thousands of plant and animal species traveled around the world. Only a few species could survive in the new destination. However, this was enough to provide us with an increasing arsenal of non-native species over the centuries. And ultimately, once they had made it into human settlements, there were always open doors and windows or cracks and crevices to get inside the buildings without having to resort to the infamous drainpipes (Box 2.2).

Box 2.2: Drainpipes, Toilets, and Other Dreadful Places
Among the most persistent rumors about the origin of spiders in buildings is the assumption that they can come up drainpipes, from plugholes of any kind, or even from the toilet bowl itself. Aren't the large, brown, hairy spiders that sometimes sit in the sink or bathtub in the morning proof enough of this horror story? Well, the spider in the sink has fallen in, not climbed up from the plughole (Sect. 5.2). And spiders that could survive in this wet underworld (with lots of feces and little oxygen) do not exist. There is also no evidence that spiders enter our homes via drains. Even if we indecently use the toilet to dispose of spiders, they will never come back.

House spiders could be brought to Europe from all regions of the world through **trade** on the one hand. On the other hand, European house spiders also made it to all parts of the world. We imagine that many of the species spread as house spiders could not survive everywhere. However, as global trade caused a strong mixing of the house spider community and there was, and always is, a new supply, we can find house spiders in almost every building in the world today. This is facilitated by the uniform basic structure of the buildings and their pleasant climate.

The definition of house spiders includes not only the fact that they occur in buildings but also that they can live and reproduce there over the long term. This concerns questions of indoor climate, addressed above, but also of nutrition (see below), requirements for a hiding place, for the web, and for egg-laying, as well as relative freedom from competition and enemies. House spiders are adaptable, undemanding, and quite voracious in all these areas.

However, there are always spiders that run into buildings from outside but cannot survive there. All over the world, you occasionally find a wolf spider (Lycosidae) in a building, but it cannot remain there for a long time. Other species of garden spiders are occasionally found as individual animals in buildings but cannot maintain themselves there due to the unfavorable living conditions. We refer to such animals as incidental findings and do not consider them further.

2.4   What Do House Spiders Live On?

Since we regularly find several spider species in houses, one must assume that there must be enough prey for house spiders inside. But what do the spiders that live with us actually eat? Most homeowners and apartment dwellers will claim that there are hardly any insects or other animals suitable as spider prey within their four walls. More precisely, they assume this or they would like it to be so. The reality, however, looks quite different.

Just as spiders always find access to our homes, many insects can easily get in as well. Windows and doors are the most obvious entry points, but gaps under closed doors or window blinds also offer important access opportunities. Furthermore, with everything we bring into our homes, **insects** can also come along as stowaways. Bouquets of flowers and potted plants are known as Trojan horses for insects. Many foodstuffs also contain insects, which we then (because of our human perspective) call food pests.

House spiders cannot afford to be picky about their food. They must try to catch and eat everything that presents itself to them. They are unspecialized (in terms of prey) and modest. In addition, they must also be able to overcome difficult or dangerous prey animals. **Woodlice** are especially common in the damp parts of buildings, but due to their exoskeleton and their ability to roll up into a ball, they are not easy prey. So anyone who can eat woodlice, in addition to their usual prey spectrum, is at an advantage.

The same applies to **ants**. These are quite defensive and can seriously injure even larger spiders. Ants often occur in considerable numbers, so they are a worthwhile prey item. Therefore, anyone who can overcome them also has an important advantage.

Ultimately, most house spiders in a building also produce quite a lot of offspring. Many spider species can produce an eggsac several times a year, each of which contains several dozen eggs. So one can easily calculate how the spider density in our home would increase explosively in a few years. But this is not the case, as we can quickly determine. **Competition** among spiders is high, and the ability to eat other spiders is strongly pronounced (Fig. 2.1). House spiders are therefore the most important regulator to keep the population density in our house constantly low. Spiders simply love to eat their neighbors. And this strong competitive pressure also claims most spiders that occasionally stray into the house, such as garden spiders.

Fig. 2.1 The long-bodied cellar spider *Pholcus phalangioides* (cellar spiders, Pholcidae) eats the common ground spider *Drassodes lapidosus* (ground spiders, Gnaphosidae). (*Photo* Jutta Ansorg)

2.5 Advantages and Disadvantages of House Spiders

After what has just been said, one of the advantages of spiders is that they catch and eat insects and other small animals (Fig. 2.2). Yes, even spiders are kept at a low population density. House spiders thus free us from woodlice, silverfish, springtails, cockroaches, ants, mosquitoes, fungus gnats, fruit flies, houseflies, pantry moths, clothes moths, larder beetles, wood beetles, and many other insects that are annoying or harmful in the house. More precisely, they do not completely eradicate these animals, because otherwise there would be nothing left to eat tomorrow, but they significantly reduce their numbers. This alone benefits or helps us.

What is our alternative if we wanted to forego this support from house spiders? With a considerable amount of hunting zeal, hygiene, washing and cleaning actions, and **insecticides**, we could also rid ourselves of a large number of insects. But few of us will succeed in keeping our home completely animal-free for a longer period of time. Moreover, the effort and cost are high; especially the commonly available insecticides are not without side effects and can have a detrimental effect on our health. So we have the choice to continue to use the free service of spiders or to accept significantly more small creatures, including those unpleasant to us.

In discussing the pros and cons of spiders in the house, sometimes the strangest things are mentioned. For example, spider webs in the house are said to **attract dirt**. If this were really the case, we would have the perfect vacuum cleaner. Spider webs would collect dirt and, with the occasional removal of the spider webs, the dirt

Fig. 2.2 Encased prey remains on the floor under a web of the long-bodied cellar spider *Pholcus phalangioides* (cellar spiders, Pholcidae). (*Photo* Jutta Ansorg)

would also be removed. That would be nice. Spider threads can hold all kinds of particles that are moved with the air stream, either as adhesive threads or through static interactions, including prey, because that's what they were made for. In a spider web, we therefore see the dirt particles that are already present in our home, freshly caught insects, and the prey remains from the last few days. The spider web simply holds up a mirror to us and shows us what is already around us anyway (Fig. 2.3).

It is also often emphasized that spiders in the house indicate a **good indoor climate**. Is this true? As explained above, house spiders are the result of a strict selection for tolerance toward the climatic conditions in our houses, which are mainly characterized by dryness. Such a dry climate is not tolerable for garden spiders, and therefore they do not survive in our buildings. The spiders that can survive in our houses, therefore, do not indicate a particularly good indoor climate, but only that they, unlike the vast majority of spider species, can also survive in our homes. By the way, an indoor climate that is too dry is not always healthy for humans either.

And then it is occasionally even claimed that spiders in the house are supposed to bring **luck**. The scientific basis (i.e., the evidence) for such a statement is unknown to us. But perhaps one should first define the term luck. We authors agree that the idea of luck and spiders is basically nonsense. But, of course, as spider researchers, we are always happy when we encounter spiders.

A serious aspect, however, is that there are people who **are afraid of spiders**. We have extensively discussed the causes of this fear of spiders and the possibilities of getting rid of it in a previous book. Through constantly trying to avoid a

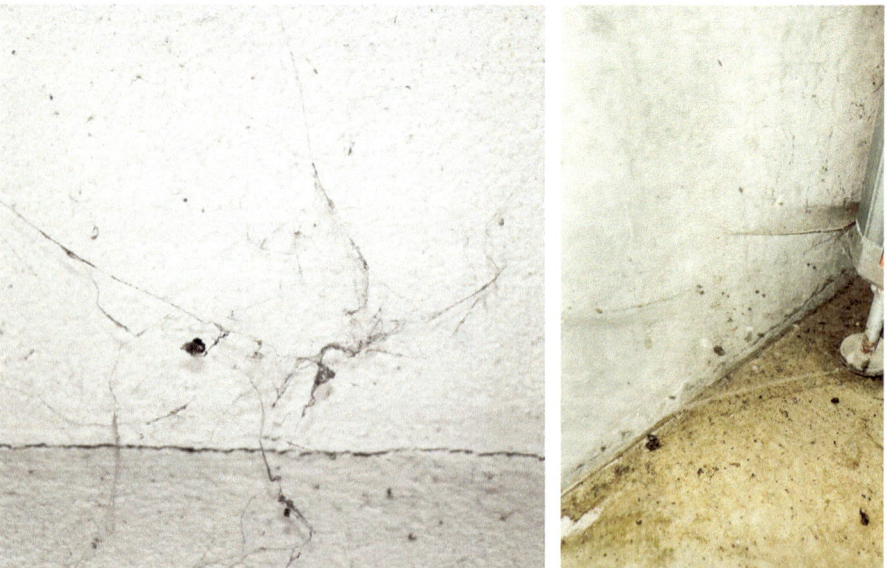

Fig. 2.3 House spider webs dirtied by dust and prey remains. (*Photo*s Jutta Ansorg (left), Ambros Hänggi (right))

confrontation with their own house spiders, such fear could even become intensified. On the other hand, house spiders, through their (harmless) presence, also offer us the opportunity to get used to these animals and overcome our fear of spiders (Box 2.3).

Box 2.3: House Spiders Help Overcome Fear of Spiders

Fear of spiders has no rational cause, as almost all spiders are harmless to humans. Moreover, they are usually very small animals but useful to us as insect predators. The vast majority of spiders cannot bite us and do not want to do so anyway. Spider venom evolved to kill insects. It is therefore produced by spiders in small quantities that are completely irrelevant to us humans. Since the production of their venom is very costly for spiders, they would never waste it on humans, whom they do not consider as potential prey.

Only a few spiders are capable of biting us due to their size and behavior. These are usually accidents where the spiders were unwittingly crushed or cornered by us. Such spider bites have usually been compared in terms of their effects to mosquito or wasp stings. Health-threatening consequences are extremely rare, and deaths have hardly occurred for decades. In contrast to spiders, the health effects (including deaths) from bee stings are many times greater.

Nevertheless, fear of spiders is widespread in Europe and is traditionally passed on in the family or in kindergarten and school. The roots of this fear of spiders go far back into the (really dark) Middle Ages and have to do with superstition, disease, death, and the devil. Today we know that the first step toward overcoming this fear is to take a closer look at spiders, i.e., to acquire factual knowledge and get used to the animals. A good start is to look closely at a spider in a picture, for example, in the first chapter of this book. Where is the front, where is the back, how many legs and eyes does a spider have, where are the spinnerets, and what does it do with them? Next, one should try to apply this knowledge to a living spider. We recommend, for example, observing a common orb-weaver near the house. Watching it build its web or catch prey is exciting and helps to foster respect for these fascinating (and not terrifying) animals. In your own home, it can be animals that live in a corner in a web. But you can also catch a spider with a glass (below is how to do it) and then calmly look at it from all sides.

Repeated viewing of the animals will lead to more knowledge about spiders. Once a house spider is in a glass, you can also keep it there without any problems for a few days. If it is, for example, the long-bodied cellar spider *Pholcus phalangioides*, which we recommend as the first spider to try, it will build a web and catch prey in a sufficiently large container (terrarium) as soon as we add something it can eat. This spider is a very rewarding object of observation. House spiders can thus help people overcome their fear of spiders through regular observation and habituation. One last tip: give the house spider in the terrarium a human name; it helps to develop a personal connection and build closeness. Why not Gertrude?

2.6   How to Control House Spiders?

There are people who are happy about spiders in the house and like to observe them. Others feel less comfortable with these subtenants and want to get rid of them (or at least some of them). We, the authors of this book, are also positive about spiders in the house, but here we still cover some methods to get rid of house spiders in a way that is compatible with both spiders and humans.

The method of choice for catching a spider on the wall involves an **empty glass** or yogurt cup and a cardboard lid. The glass is slowly placed over the spider, then the cardboard lid is inserted between the glass and the wall, and the spider is trapped. Alternatively, you can also use a spider catcher (available in shops), a grab-arm-like device that allows you to catch the spider at a distance and release it outside.

The method probably used most often involves the **vacuum cleaner**. This method is deadly for spiders, as they are crushed in the device by the strong suction. Spiders, therefore, do not survive this method, so they cannot climb out of the vacuum cleaner at night, as is occasionally claimed. We reject this method because there is no reason to kill the animals.

A range of sprays is offered in stores that are supposed to either kill or repel spiders. The repelling is usually based on **essential oils**, and as far as there are useful scientific tests for this, their effectiveness is not convincing. Various **insecticides** are offered to kill spiders, all of which have in common the fact that they are quite effective. These are usually substances that act on the nervous system and also negatively affect the people, dogs, and cats living in our home. Because of their toxicity, we therefore reject these methods.

But where do you release captured spiders? Simply setting spiders outside the front door is likely to do little, as the spiders will quickly reappear in your own house. Whether it makes sense to release them in front of your neighbor's house may depend on your relationship with your neighbor. Since real house spiders cannot survive on a lawn, in a city park, or in a potato field on the outskirts of the city, the best idea is probably to release house spiders a few houses further away at basement windows or next to industrial buildings, sheds, or warehouses.

2.7   Spiders Outside the House

Anyone who deals with buildings, and the spiders associated with them, quickly realizes that there are also a number of spider species outside the buildings that can be found there with great regularity. The exterior of our buildings is more microclimatically related to the surroundings than the interior of the building, which means that persistent dryness and frost-free conditions are not as important there. The masonry of a building is often quite structurally poor, heats up strongly in the sunshine and is dry, cools down just as strongly at night, and becomes very wet in the rain. Rock-dwelling spider species can live with these temperature extremes and thus find an excellent **substitute habitat** on the outside of buildings.

Window cavities, gaps in door and window frames, or window blinds represent excellent substitute habitats for spider species that otherwise live under bark or in holes and cracks in old tree trunks. In a similar way, the space under a protruding roof, in passages, or between closely standing buildings is often used by web-building species that need stable support structures for their large webs.

The spiders that we find on the outside of buildings often come from the spider community that occurs in the wider surroundings of the buildings. Worldwide, more species are found outside of buildings than inside. Nevertheless, in this book, we will also introduce the most common species worldwide that can be found on the outside of houses.

A survey of 100 spider experts from various countries and an analysis of the specialist literature provided a wealth of information on the occurrence of spiders in buildings. Based on this, the most common spider species in buildings throughout the world were selected, each of which is presented here with a global distribution map.

Everything started with an idea and many emails to international colleagues. The central question was, "Which spider species live in your buildings?" All of these specialists, of course, know the spiders that occur in their houses and have willingly provided us with answers. In total, we received about 100 responses to our emails. The names of these colleagues are listed in our acknowledgments. In addition, the scientific literature was reviewed, and 30 publications dealing with spiders in buildings throughout the world were analyzed. Last but not least, we were able to query four databases in which spider findings from homes and buildings were particularly noted.

The collection and evaluation of all this information in our house spider database resulted in 2029 records of spiders in buildings, belonging to 642 species. These two numbers alone suggest that, in addition to **real house spiders**, which are found repeatedly in buildings worldwide, many **incidental findings** of such spiders are also included. These actually live outside of buildings. We therefore focused on frequent mentions, i.e., the approximately 50 species or species groups that can be found regularly and on several continents within houses. These are the species discussed here, which we refer to as real house spiders.

In addition to a numerical evaluation, we also brought our own personal **experience** to bear. Wolf spiders (Lycosidae), belonging to different species, keep appearing in buildings all over the world. However, for us, wolf spiders are typical incidental findings of spiders that can indeed frequently enter homes and buildings but cannot live there for a longer period of time. The situation is similar for crab spiders (Thomisidae). Crab spiders are occasionally found on or in buildings all over the world, but there is no specific species that can complete its life cycle in or on a building. We have therefore also classified crab spiders as incidental findings.

On the other hand, we were disappointed that not more pirate spiders (Mimetidae) were detected in buildings. These small spiders are not attracted to our buildings as

© Association for the Promotion of Spider Research 2024
W. Nentwig et al., *House Spiders - Worldwide*,
https://doi.org/10.1007/978-3-031-70448-2_3

a habitat but particularly like to eat cobweb spiders (Theridiidae). We therefore regularly find pirate spiders in the intricate webs of cobweb spiders, which are common in our buildings but are often overlooked by experts. Thus, in this book, we did not consider wolf spiders (Lycosidae) and crab spiders (Thomisidae), but we did consider pirate spiders (Mimetidae).

We show a **distribution map** for each spider species presented here. The following should be noted: in practice, the precise worldwide distribution of a spider species is never known accurately. A distribution map is therefore always only an approximation of reality, which primarily reflects the insufficient state of research. We use three representations in the maps.

1. A country filled in color indicates the presence of a spider species. We chose this representation when there are many records from this country. However, for very large countries that possibly extend over several climate zones (such as Canada, the United States, Chile, Russia, and Australia), this does not mean that the species occurs everywhere in the country. Usually, we then represent the presence of a species with points in such large countries.
2. A circle encloses a group of countries or a region where the species occurs. However, within this circle, there may well be areas where it is currently unknown whether the species actually exists.
3. Points represent geographically limited occurrences in a country. We have mostly used points when nothing more is known about a species for a country except for a single record.

It is important to note that color-filled countries do not always represent the original **distribution area** of a species. On the other hand, points in isolated occurrences represent countries where this species has been introduced. Often, the origin and distribution history of the species are not well known. More details can be found in the text or in the cited scientific literature, if known.

In addition to the data from our survey of international experts, we have taken information for the distribution maps from the scientific literature, which is also cited for the respective species. Furthermore, we used information from the World Spider Catalog, as well as the literature cited there and data from the global biodiversity database GBIF (2023).

The coloring of the distribution maps was chosen so that they can also be easily distinguished by color-blind people.

Identification Keys for Spiders in the Home and on the House

4

In this chapter, we offer various identification aids to determine the house spiders discussed here. We will not go to the species level but will be able to resolve their family affiliation. Such identification keys differ in their difficulty depending on the spider species and your prior knowledge. For this reason, in addition to a classic dichotomous identification key, where you always decide step by step between two characteristics, we also offer identification aids based on how commonly the spiders occur and some of their particularly striking features.

Anyone who wants to use this book would naturally expect an identification key for the most common species in and around the house. That is what we also planned. However, when working on this chapter, we quickly realized that a well-functioning identification key is anything but trivial, especially since we want to present it as simply as possible. We would like to emphasize one very important point: the identification key presented here only refers to the spiders that regularly occur in or on houses. In the natural environment, there are many more spider species, which, of course, do not appear in the key. On the other hand, there are also species out there that are very similar to the spiders in the house, so they would seem to appear in the key but are actually different species. Our approach toward identifying house spiders should therefore only be used for these spiders found in houses.

4.1   How Does This Identification Key Work?

Identification keys work in such a way that a series of **characteristics** are queried one after another. In question 1, for example, you are asked whether characteristic A or characteristic B is present. The key then continues with either number 2 or 3. Useful characteristics for spiders are, in practice, almost exclusively anatomical. In other words, they concern certain body parts and structures. Since spiders are mostly small, identifying them is often a tedious task because the animal is constantly scrutinized for special, but tiny, structures that often (in a key with only two alternatives) are not present.

© Association for the Promotion of Spider Research 2024
W. Nentwig et al., *House Spiders - Worldwide*,
https://doi.org/10.1007/978-3-031-70448-2_4

To avoid this fundamental difficulty, we offer here a different approach, or more precisely, several different approaches. We generally refrain from including very small structures and especially characteristics in the area of the spiders' sexual organs. For exact species identification, these must, of course, be examined too, but this requires a binocular microscope. Most people do not normally own such equipment, as it is quite expensive. For the characteristics used here, such as the eyes, body shape, and spinnerets, a good magnifying glass is sufficient, the purchase of which is highly recommended for observing many small creatures and not just spiders (Box 4.1).

Box 4.1: The Classification System for Spiders
We do not want to delve into the complex and also often contradictory systematics of spiders here and have therefore reduced the classification system used below to a minimum. The spider **order** (Araneae) currently includes over 52,000 species, according to the World Spider Catalog. Closely related species are grouped into **families**, of which we currently recognize over 130. Within a family, there is further subdivision into **genera** (of which we currently recognize over 4300), such that each genus contains the closest related **species**. To illustrate this with an example, the barn funnel weaver *Tegenaria domestica* belongs to the genus *Tegenaria*. This genus contains, besides the species *domestica*, another 128 additional species. *Tegenaria* belongs, along with 95 other genera, to the funnel-web spider family (Agelenidae). Latin genus and species names are always written in italics to indicate that they are genus or species names. Family names are never written in italics. In animals, they can be recognized by the fact that they always end in -idae (as in our example of Agelenidae). In English, the family names can also be linguistically adapted, so you can also refer to them as agelenids.

To be able to easily recognize and identify the spider species presented here, it is often necessary to catch the spider. Once in a glass, the animal can then be observed in peace. We would also like to introduce a simply constructed container in which a spider can be held alive so that it can be easily observed without becoming injured (Fig. 4.1).

Alternatively, a captured spider can also be placed into a clean and transparent plastic bag. If you now carefully lay the bag flat, without crushing the spider, you can restrict its movement and view it from all sides with a magnifying glass.

Our suggestion for a "simple" identification of spiders in and around the house leads in this chapter to the corresponding family. In the following chapters, individual families are presented, and based on the illustrations, you can then decide on a particular species.

Fig. 4.1 Homemade observation container for spiders, which allows the living spider to be held without injury. With a hand magnifier, the immobile spider can be easily examined. For this, only three plastic coffee cream containers, foam, and transparent plastic film are needed. The bottom is cut out of two cups with a sharp knife, and a piece of transparent film is stretched between them. A piece of foam is glued to the third cup and pushed into the other two cups with the film. A spider caught between them cannot move and can be viewed through the film with a magnifying glass. (*Photo* Wolfgang Nentwig)

However, the "simple" identification of the spider family depends heavily on prior knowledge. We therefore present a procedure that follows the principle of set theory, namely total set and subset: of the 23 families that we consider house spiders and discuss here, some are difficult to recognize, while others are not. If you first look at the families with striking features, there is a high probability of already achieving a hit, or at least some families can be eliminated.

As a first method, we list here some particularly striking features and mention the spider families to which these apply. Often, this already helps to identify an animal. Then, as a second method, we will use these features (and some additional ones) to facilitate identification via a classic dichotomous key, as explained above. In general, it may also be helpful to know which spider families are particularly likely to be found in buildings. This information can be found in Box 4.2. Of course, it also makes sense to compare a spider to be identified with the photos in the later part of this book. This simplest identification technique of image comparison, which is the standard method in identification books for butterflies, is more difficult with spiders but can still be helpful.

Box 4.2: The Most Common Spider Families in the House

Why not use the frequency with which species or families occur to orient yourself? It is quite likely that one or more of the world's most common house spiders also occur in your home. Therefore, we have calculated the most common spider families according to the house spider database described in Chap. 3 as a percentage of all mentions and listed them here. The most common species were already discussed in Table 2.1. Accordingly, these are the most common families of house spiders:

15 % of all house spider records are **cobweb spiders** (Theridiidae). These include many species of web-building spiders, especially in the genera *Steatoda* and *Parasteatoda* as well as *Latrodectus* (Chap. 25).

13 % of all house spider records are **jumping spiders** (Salticidae), including many different species, most commonly from the genera *Hasarius* and *Menemerus*, which run around on walls (Chap. 19).

11 % are **cellar spiders** (Pholcidae) with many species, mostly in rather messy webs, the most common being *Pholcus phalangioides* (Chap. 18).

7 % of all house spider records are **orb-weavers** (Araneidae), represented by many different species mostly outside of buildings, and most commonly from the genera *Araneus* and *Zygiella* (Chap. 7).

6 % are **funnel-web spiders** (Agelenidae), especially species of the genera *Eratigena* and *Tegenaria* (Chap. 5).

4 % are **ground spiders** (Gnaphosidae), especially species of the genus *Scotophaeus* (Chap. 13).

4 % are **spitting spiders** (Scytodidae), especially *Scytodes* species, the most common of which is *Scytodes thoracica* (Chap. 20).

3 % are **disk web spiders** (Oecobiidae), with many *Oecobius* species (Chap. 16).

As a general note, spiders molt about five to ten times until they reach adulthood, with smaller species less and larger species more. For example, a young garden cross spider at the third or fourth stage looks quite similar to the adult animal but is significantly smaller. **Size specifications** for spiders (in general and especially here in this book) refer only to adult animals and are usually given separately for both sexes, as males are often slightly smaller than females. The **body length** of a spider extends from the front edge of the front body to the rear edge of the hind body, without adding mouthparts, spinnerets, or legs.

If you only want to orient yourself by considering the spider's body size, you need to exclude young animals. But how do you know if a spider is already an adult

or still immature? By their sexual organs! Adult males have two thickened append-
ages (pedipalps) at the front of the body, which look a bit like boxing gloves at the
ends and function almost like syringes (Fig. 1.2). With these, the sperm is sucked up
and transferred into the female. Adult females in most of the families treated here
exhibit a hard plate on their underside called the epigyne (Fig. 1.1). This serves the
male as an insertion aid for his palp and allows him to hook onto the female.
Unfortunately, in some spider families, this epigyne is poorly pronounced or even
missing. As a rule of thumb, if a spider is as big as we indicate here in the presentation of the species, it is usually an adult.

4.2   Distinctive Identification Features in House Spiders

At first glance, certain features of a spider can stand out, allowing for quick and easy
identification of the family. A comparison with the spider species presented in the
following chapters allows you to quickly check whether your initial assumption was
correct (Box 4.3).

Distinctive Legs
All legs very long and thin, like a harvestman:
• Cellar spiders (Pholcidae) (Chap. 18)
Front legs crab-like and oriented sideways:
• Giant crab spiders (Sparassidae) (Chap. 24)
• Flatties (Selenopidae) (Chap. 22)
Front legs long and spiky, rather small spiders:
• Pirate spiders (Mimetidae) (Chap. 15)
Long front legs bearing tufts of hair:
• Hackled orb-weavers (Uloboridae) (Chap. 26)

Distinctive Eyes
Six eyes only:
• Recluse spiders (Sicariidae) (Chap. 23)
• Tube-dwelling spiders (Segestriidae) (Chap. 21)
• Woodlouse hunters (Dysderidae) (Chap. 11)
• Spitting spiders (Scytodidae) (Chap. 20)
• Cellar spiders (some Pholcidae) (Chap. 18)
• Goblin spiders (Oonopidae) (Chap. 17)
Front middle eyes are noticeably larger than all other eyes:
• Jumping spiders (Salticidae) (Chap. 19)
Eight eyes on a common eye mound:
• Crevice weavers (Filistatidae) (Chap. 12)

Box 4.3: "What's Its Name?" or the Trouble with Names
The names for spider species, genera, or even families in the respective national languages, be it in English, German, Hindi, or Chinese, are so-called **common names** or **popular names**. In most countries, these names are not assigned according to any official system, and there is no committee that assigns these names. A remarkable exception is the American Arachnological Society, which has a Common Names Committee. This society establishes common names for North America. However, the same spider genus or species will probably have completely different common names in different regions. Moreover, new common names arise in the public and change through language use.

Often, spiders are named after external features (like the zebra jumping spider) or after striking behaviors or places where they live (like the cellar spider). A major disadvantage of these common names is that they are often regionally different and also not unique. For example, the term "daddy long-legs" can refer to a harvestman (Opiliones, not closely related to spiders), a crane fly (Tipulidae, a harmless relative of mosquitoes), or even a cellar spider (Pholcidae, a "true" spider). Another disadvantage is that less noticeable species are not named by the general population at all. Also, there is not just one cellar spider, but several species all over the place, which are usually not assigned different common names. This is the reason why sometimes no popular name for a species is mentioned in this book. We do not want to invent new names because this might just add to confusion.

For **scientific purposes**, common names are also not especially useful. Therefore, the Swedish naturalist Linnaeus proposed in 1758 a system for a two-part naming with Latin names for the genus and the species. This system has been maintained until today and is now regulated by an international commission for nomenclature. Every known animal and plant species, and thus every spider species, has since acquired a unique scientific name. However, this does not mean that these names cannot sometimes be changed according to established rules, as explained in Box 5.1 on the occasion of the *Tegenaria* earthquake.

Front or Hind Body with Humps
- Hackled orb-weavers (Uloboridae) (hind body pointed upwards and tapering) (Chap. 26)
- Spitting spiders (Scytodidae) (front body as high as the hind body, never flat) (Chap. 20)
- Pirate spiders (Mimetidae) (hind body with two or four humps) (Chap. 15)

Very Small Spiders (Body Length up to 3 mm)
- Sheet-weavers (Linyphiidae) (some species) (Chap. 14)
- Mesh-weavers (Dictynidae) (some species) (Chap. 10)

- Disk web spiders (Oecobiidae) (Chap. 16)
- Cellar spiders (Pholcidae) (*Spermophora* and others) (Chap. 18)
- Goblin spiders (Oonopidae) (Chap. 17)

Very Large Spiders (Body Length More Than 15 mm)
- False wolf spiders (Zoropsidae) (up to 19 mm) (Chap. 27)
- Ground spiders (Gnaphosidae) (*Herpyllus* up to 17 mm) (Chap. 13)
- Orb-weavers (Araneidae) (some species up to 40 mm) (Chap. 7)
- Giant crab spiders (Sparassidae) (up to 36 mm) (Chap. 24)
- Funnel-web spiders (Agelenidae) (some species up to 16 mm) (Chap. 5)

Appearance: "Large, Brown, Hairy"
- False wolf spiders (Zoropsidae) (Chap. 27)
- Funnel-web spiders (Agelenidae) (Chap. 5)

Noticeably Colorful Spiders
- Cobweb spiders (Theridiidae) (*Latrodectus* species with red, orange, brown, or yellowish color pattern on a black-glossy background) (Chap. 25)
- Cobweb spiders (Theridiidae) (*Nesticodes rufipes* with reddish front body and legs) (Chap. 25)
- Orb-weavers (Araneidae) (often colorful patterns) (Chap. 7)
- Woodlouse hunters (Dysderidae) (front body shiny reddish-brown) (Chap. 11)
- Spitting spiders (Scytodidae) (yellowish to reddish-black pattern) (Chap. 20)
- Jumping spiders (Salticidae) (some species with colorful patterns) (Chap. 19)

Hind Body Velvety
- Hackled mesh-weavers (Amaurobiidae) (Chap. 6)
- Crevice weavers (Filistatidae) (Chap. 12)

Webs
Orb-webs (see illustrations in Chap. 7):
- Hackled orb-weavers (Uloboridae) (small, with cribellate catching threads) (Chap. 26)
- Orb-weavers (Araneidae) (large, threads with adhesive droplets) (Chap. 7)
Funnel-webs (Fig. 5.1):
- Inter-tidal spiders (Desidae) (messy, with cribellate catching threads, Fig. 9.1)
- Funnel-web spiders (Agelenidae) (large, classic funnel-webs, Fig. 5.1)
Tube web with only a few catching threads distant from the tube edge:
- Hackled mesh-weavers (Amaurobiidae) (with cribellate catching threads, Fig. 6.2)
- Tube-dwelling spiders (Segestriidae) (with cribellate catching threads, Fig. 21.1)
- Crevice weavers (Filistatidae) (with cribellate catching threads, Fig. 12.1)
Irregular, three-dimensional webs:
- Sheet-weavers (Linyphiidae) (sometimes with noticeable web sheet) (Chap. 14)

- Cobweb spiders (Theridiidae) (very messy arrangement, sometimes with some sticky droplets near the ground) (Chap. 25)
- Mesh-weavers (Dictynidae) (irregular web with cribellate catching threads, Fig. 10.2)
- Cellar spiders (Pholcidae) (very messy arrangement, extremely fine and therefore barely visible) (Chap. 18)

Spiders Without Webs for Prey Capture
- Yellow sac spiders (Cheiracanthiidae) (Chap. 8)
- Recluse spiders (Sicariidae) (nocturnal) (Chap. 23)
- False wolf spiders (Zoropsidae) (nocturnal) (Chap. 27)
- Ground spiders (Gnaphosidae) (nocturnal) (Chap. 13)
- Giant crab spiders (Sparassidae) (nocturnal) (Chap. 24)
- Woodlouse hunters (Dysderidae) (nocturnal) (Chap. 11)
- Spitting spiders (Scytodidae) (nocturnal) (Chap. 20)
- Jumping spiders (Salticidae) (diurnal) (Chap. 19)
- Flatties (Selenopidae) (nocturnal) (Chap. 22)
- Goblin spiders (Oonopidae) (Chap. 17)

Notable Behaviors
- Giant crab spiders (Sparassidae) (can run very fast and also sideways) (Chap. 24)
- Disk web spiders (Oecobiidae) (run quickly out of their tent-like web retreat) (Chap. 16)
- Spitting spiders (Scytodidae) (the cocoon is carried by the spider in the mouthparts) (Chap. 20)
- Jumping spiders (Salticidae) (very active, can jump well) (Chap. 19)
- Flatties (Selenopidae) (can run very fast and also sideways) (Chap. 22)
- Cellar spiders (Pholcidae) (noticeable rapid swinging back and forth in the web when threatened; cocoon is carried by the spider in the mouthparts) (Chap. 18)

4.3   Dichotomous Identification Key for House Spiders

A preliminary note on spider eyes: depending on whether you look at the eyes from the front or from above, their shape and position relative to each other can look different. Therefore, in the following eye drawings, we usually provide both views: from above (left illustration) and from the front (right image).

1a	All legs are very thin and very long, reaching more than twice the body length, sometimes three to four times. This gives the animals a harvestman-like appearance. In contrast to harvestmen, however, spiders have a two-part body. The web is irregular, without a retreat or retracted silk cover, often hard to recognize, can be connected with neighboring webs, and is often dirty. When the web is touched or an attempt is made to catch the animal, it performs rapid, oscillating movements (trembling), so that they blur before our eyes. Body length: 2–10 mm. Six eyes in two groups of three (*Spermophora*) or eight eyes with very small front middle eyes, as shown in Fig. 4.2. The cocoon is held in the mouthparts. **Cellar spiders (Pholcidae)** (*Artema, Crossopriza, Holocnemus, Pholcus, Psilochorus, Smeringopus, Spermophora*) (Chap. 18)  **Fig. 4.2** The eye arrangement of cellar spiders (Pholcidae) viewed from above, with six eyes (**left**) and with eight eyes (**right**). (*Figure* Miriam Frutiger)
1b	Different 2
2a	Front and hind body with a striking yellowish to reddish-black pattern, front body noticeably domed, legs thin and annulated, eyes arranged in three pairs with the middle eyes shifted forward (Fig. 4.3). Body length: 4–11 mm. The cocoon is held in the chelicerae. **Spitting spiders (Scytodidae)** (*Scytodes*) (Chap. 20) 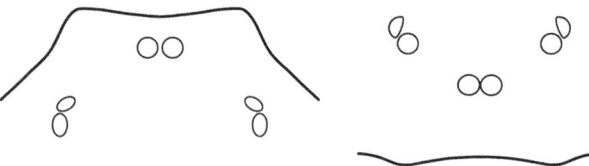 **Fig. 4.3** The eye arrangement of spitting spiders (Scytodidae) viewed from above (**left**) and from the front (**right**). (*Figure* Miriam Frutiger)
2b	Different 3
3a	Noticeably large front middle eyes, with the rear middle eyes very small (Fig. 4.4). Compact body structure with short, robust legs. Excellent jumping ability. Small to medium-sized spiders, body length: 3–14 mm. **Jumping spiders (Salticidae)** (*Hasarius, Marpissa, Menemerus, Phidippus, Plexippus, Pseudeuophrys, Salticus*) (Chap. 19) 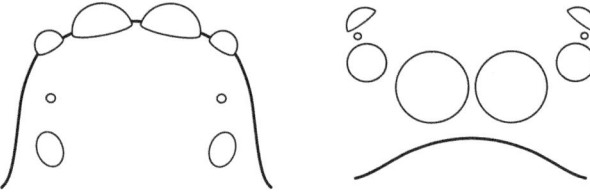 **Fig. 4.4** The eye arrangement of jumping spiders (Salticidae) viewed from above (**left**) and from the front (**right**). (*Figure* Miriam Frutiger)
3b	Different 4

4a	First pair of legs is at least one and a half times as long as the rest, not thickened but with thin spines or with dark tufts of hair. **5**
4b	Different **6**
5a	First pair of legs is noticeably strong and long, with dark tufts of hair, and the hind body with two humps. Webs are small, cribellate orb webs (with crimped wool on the spiral threads). Often, there are several next to each other, frequently on houseplants or other structures in the house. Eight small eyes in two semicircles, the front middle eyes slightly larger, as shown in Fig. 4.5, body length: 3–8 mm.

Hackled orb-weavers (Uloboridae) (Uloborus) (Chap. 26)

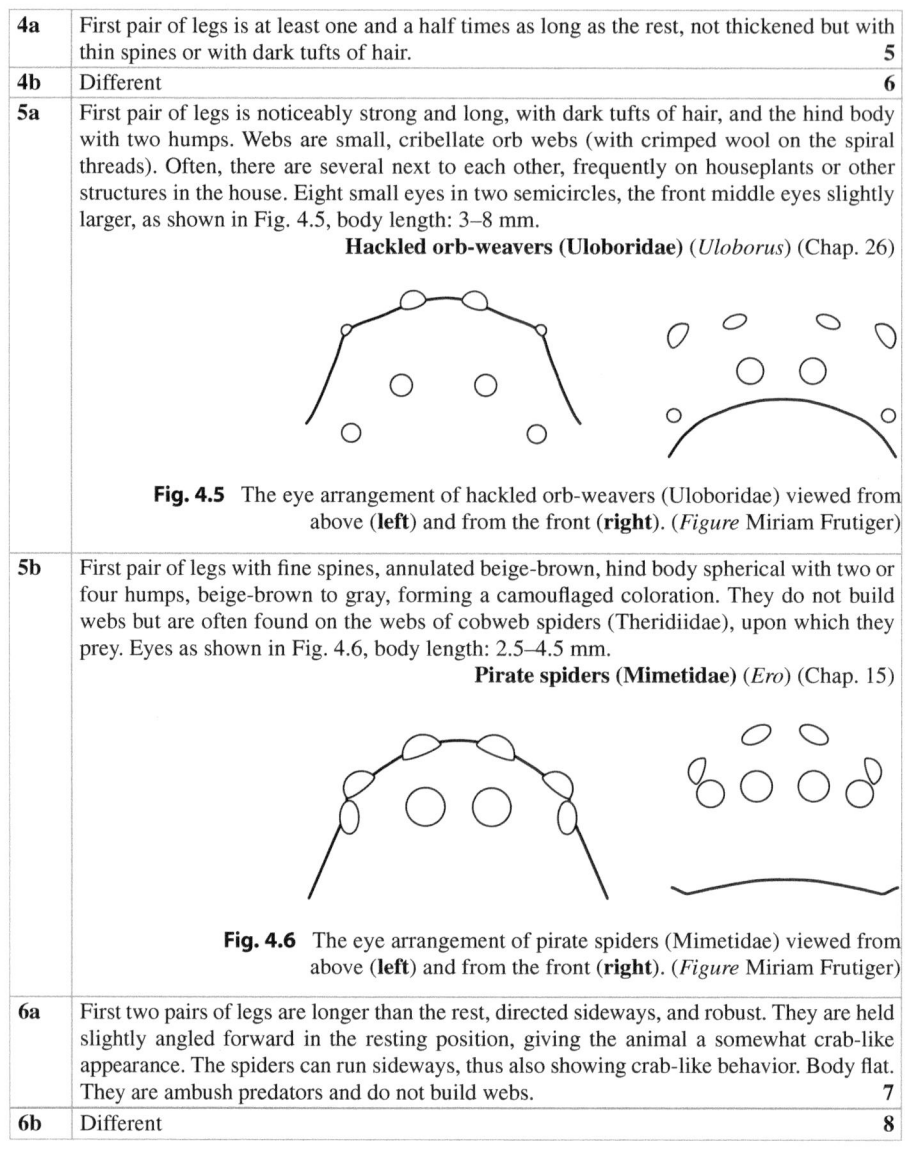

Fig. 4.5   The eye arrangement of hackled orb-weavers (Uloboridae) viewed from above (left) and from the front (right). (Figure Miriam Frutiger)

5b	First pair of legs with fine spines, annulated beige-brown, hind body spherical with two or four humps, beige-brown to gray, forming a camouflaged coloration. They do not build webs but are often found on the webs of cobweb spiders (Theridiidae), upon which they prey. Eyes as shown in Fig. 4.6, body length: 2.5–4.5 mm.

Pirate spiders (Mimetidae) (Ero) (Chap. 15)

Fig. 4.6   The eye arrangement of pirate spiders (Mimetidae) viewed from above (left) and from the front (right). (Figure Miriam Frutiger)

6a	First two pairs of legs are longer than the rest, directed sideways, and robust. They are held slightly angled forward in the resting position, giving the animal a somewhat crab-like appearance. The spiders can run sideways, thus also showing crab-like behavior. Body flat. They are ambush predators and do not build webs. **7**
6b	Different **8**

7a	Body length 15–36 mm, with long legs that permit a noticeably large leg span. The front middle eyes are smaller than the other eyes (Fig. 4.7). Nocturnal ambush predator, often on walls, on the ceiling, or under the roof. **Giant crab spiders (Sparassidae)** (*Heteropoda*) (Chap. 24) 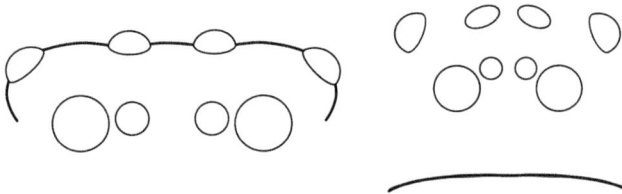 **Fig. 4.7** The eye arrangement of giant crab spiders (Sparassidae) viewed from above (**left**) and from the front (**right**). (*Figure* Miriam Frutiger)
7b	Body length: 6–23 mm. The eyes are approximately in a row, with the rear lateral eyes significantly enlarged (Fig. 4.8). The front body and hind body are flat, and the body is mostly brown or with gray tones, camouflaged and bark-colored. In nature, often on tree bark; in the house, on walls and ceilings. Nocturnal. So far, only found in tropical and subtropical parts of the world. **Flatties (Selenopidae)** (*Selenops*) (Chap. 22) 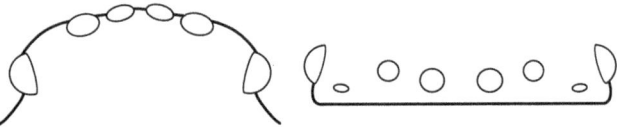 **Fig. 4.8** The eye arrangement of flatties (Selenopidae) viewed from above (**left**) and from the front (**right**). (*Figure* Miriam Frutiger)
8a	Six eyes — 9
8b	Eight eyes. Some eyes are sometimes very small or hidden between hairs. — 12
9a	The six eyes are arranged in three groups (Fig. 4.9 or 4.10) — 10
9b	The six eyes are arranged in one group (Fig. 4.11 or 4.12) — 11
10a	The front body and hind body are narrowly cylindrical, the front body is brownish to blackish, and the hind body has a chevron pattern, velvety. Six eyes in three pairs as shown in Fig. 4.9, body length: 7–15 mm. Animals sit in a tubular retreat and wait for prey, the first three pairs of legs facing forward. From the edge of the retreat, a few radial, cribellate threads lead away. **Tube-dwelling spiders (Segestriidae)** (*Segestria*) (Chap. 21)  **Fig. 4.9** The eye arrangement of tube-dwelling spiders (Segestriidae) viewed from above (**left**) and from the front (**right**). (*Figure* Miriam Frutiger)

10b	The front body is roundish, flat, with slightly raised eye part, coloration is mostly brownish, and the hind body is velvety. Eyes are arranged in three pairs, the middle eyes at about the same level as the lateral eyes (Fig. 4.10). Body length: 7–12 mm. During the day, the animals hide in nooks and crannies; at night, they roam around or wait for prey as ambush predators.

Recluse spiders (Sicariidae) (*Loxosceles*) (Chap. 23)

Fig. 4.10 The eye arrangement of recluse spiders (Sicariidae) viewed from above (**left**) and from the front (**right**). (*Figure* Miriam Frutiger)

11a	Body length 2 mm, with legs mostly yellow-orange, and front body noticeably round. Eyes about the same size, closely adjacent (Fig. 4.11). They do not build webs.

Goblin spiders (Oonopidae) (*Ischnothyreus, Oonops, Tapinesthis, Triaeris*) (Chap. 17)

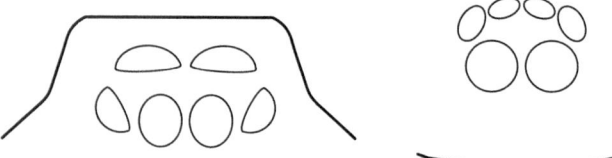

Fig. 4.11 The eye arrangement of goblin spiders (Oonopidae) viewed from above (**left**) and from the front (**right**). (*Figure* Miriam Frutiger)

11b	Body length: 10–15 mm. The front body is elongated and reddish-brown, with noticeably robust, reddish-brown chelicerae. The hind body is largely hairless, velvety, or shiny. Six eyes in a compact group as shown in Fig. 4.12. The hind body is often light in color, without chevrons. Nocturnal.

Woodlouse hunters (Dysderidae) (*Dysdera*) (Chap. 11)

Fig. 4.12 The eye arrangement of woodlouse hunters (Dysderidae) viewed from above (**left**) and from the front (**right**). (*Figure* Miriam Frutiger)

12a	Eight eyes are arranged in a compact field on an eye mound (Fig. 4.13). The front body is flat, roundish, and longer than wide, with a darker triangular field around the eyes; spiders are brownish, and hind body without chevrons appears velvety. Body length: 7–14 mm. **Crevice weavers (Filistatidae)** (*Filistata, Kukulcania*) (Chap. 12) 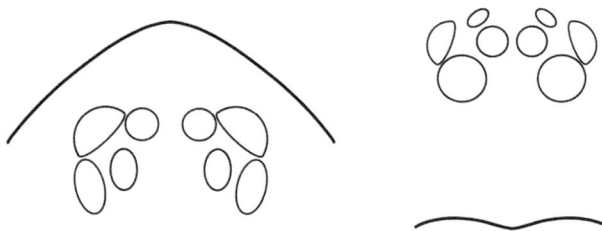 **Fig. 4.13** The eye arrangement of crevice weavers (Filistatidae) viewed from above (**left**) and from the front (**right**). (*Figure* Miriam Frutiger)
12b	Different 13
13a	Small spiders of 2–2.5 mm body length, which stand out due to their round front body with a striking pattern. All eyes are relatively close together, with the rear middle eyes being oval (Fig. 4.14). Legs are short and robust. Instead of a tube-like retreat, they build a tent-like, often hexagonal web with several exits. The spiders run quickly around their prey and wrap it with a cribellate thread. **Disk web spiders (Oecobiidae)** (*Oecobius*) (Chap. 16)  **Fig. 4.14** The eye arrangement of disk web spiders (Oecobiidae) viewed from above (**left**) and from the front (**right**). (*Figure* Miriam Frutiger)
13b	Different 14
14a	"Large, brown, hairy": 6–19 mm large animals, noticeably hairy, as nocturnal ambush predators on the wall, in a funnel web, or running around, often caught in sinks and bathtubs. 15
14b	Different 16

15a	Large funnel web; the spider usually sits at the entrance of the tubular retreat. Wandering males (rarely also females) are often caught in sinks or bathtubs. Body length: 6–16 mm; spiders are usually brownish-colored and have chevrons on the hind body. Long, highly mobile spinnerets. Eight almost equal-sized eyes; the side eyes touch each other (see Fig. 4.15). **Funnel-web spiders (Agelenidae)** (*Agelenopsis, Eratigena, Tegenaria*) (Chap. 5)  **Fig. 4.15** The eye arrangement of funnel-web spiders (Agelenidae) viewed from above (**left**) and from the front (**right**). (*Figure* Miriam Frutiger)
15b	Large spiders of 10–19 mm body length, gray-brown to reddish-brown, with a striking pattern on the hind body. All eyes are medium-sized; only the front middle eyes are slightly smaller (Fig. 4.16). Nocturnal, climbs inside and outside on walls. **False wolf spiders (Zoropsidae)** (*Zoropsis*) (Chap. 27) 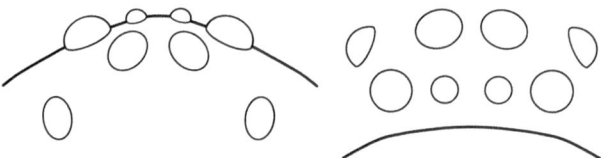 **Fig. 4.16** The eye arrangement of false wolf spiders (Zoropsidae) viewed from above (**left**) and from the front (**right**). (*Figure* Miriam Frutiger)
16a	Hind body velvety. Areas surrounding the retreat are covered with cribellate-catching threads, creating a funnel-shaped web or an extensive catching "blanket." Eight approximately equal-sized eyes are arranged in an almost oval circle; the side eyes often lie somewhat closer together (see Fig. 4.17). The front body is mostly brownish, and the hind body with dark, sometimes hard-to-recognize spots. Body length: 5–12 mm, distributed worldwide. **Hackled mesh-weavers (Amaurobiidae)** (*Amaurobius*) (Chap. 6)  **Fig. 4.17** The eye arrangement of hackled mesh-weavers (Amaurobiidae) viewed from above (**left**) and from the front (**right**). (*Figure* Miriam Frutiger)
16b	Different **17**

| 17a | Orb-weaving spider, sitting in the orb web or in a side retreat connected to the web via a thread. Medium to very large species with body lengths of 4–40 mm, often colorfully conspicuous. Eight eyes; the side eyes are particularly small, and the larger middle eyes form a square (see Fig. 4.18). Robust, not especially long legs, often with short, strong thorns; the animals cannot walk well outside of their orb web.
Orb-weavers (Araneidae) (*Araneus, Argiope, Larinioides, Nephila, Nephilengys, Nephilingis, Nuctenea, Trichonephila, Zygiella*) (Chap. 7) |

Fig. 4.18 The eye arrangement of orb-weavers (Araneidae) viewed from above (**left**) and from the front (**right**). (*Figure* Miriam Frutiger)

17b	Different	18
18a	Spiders in seemingly disordered webs or cocoons, without recognizable structure; they only leave their surroundings when they are strongly disturbed.	19
18b	Spiders without a catching web, freely roaming, active during the day or night	22
19a	Eight approximately equal-sized eyes are arranged in an oval, with the side eyes close together (Fig. 4.19). The hind body is roundish. Body length: 2.5–4 mm. Dwelling tube usually in plaster or in cracks of exterior walls, often occurring either in individual-rich colonies (*Brigittea*) or on their own (*Dictyna*). **Mesh-weavers (Dictynidae)** (*Brigittea, Dictyna*) Chap. 10)	

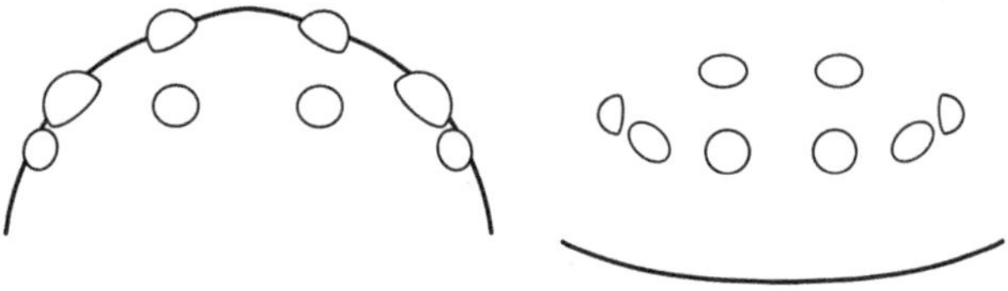

Fig. 4.19 The eye arrangement of mesh-weavers (Dictynidae) viewed from above (**left**) and from the front (**right**). (*Figure* Miriam Frutiger)

19b	Different	20
20a	Spider lives in a large funnel web, the front body is gray to black, and the hind body has chevrons. Body length: 11–15 mm. Lateral eyes and anterior median eyes each in groups together, the posterior median eyes the smallest (see Fig. 4.20). Originally limited to Australia, now introduced to New Zealand, Japan, North and South America, South Africa, and Germany. **Intertidal spiders (Desidae)** (*Badumna*) (Chap. 9)	

Fig. 4.20 The eye arrangement of intertidal spiders (Desidae) viewed from above (**left**) and from the front (**right**). (*Figure* Miriam Frutiger)

| 20b | Different | 21 |

21a	Abdomen spherical, legs only covered with short hairs, eight eyes as shown in Fig. 4.21, body length: 3–14 mm. **Cobweb spiders (Theridiidae)** (*Latrodectus, Nesticodes rufipes, Parasteatoda, Steatoda, Theridion*) (Chap. 25)  **Fig. 4.21** The eye arrangement of cobweb spiders (Theridiidae) viewed from above (**left**) and from the front (**right**). (*Figure* Miriam Frutiger)
21b	Abdomen elongated, legs with short hairs and long bristles, eight eyes as shown in Fig. 4.22, body length: 2–7 mm. **Sheet-weavers (Linyphiidae)** (*Lepthyphantes leprosus*) (Chap. 14)  **Fig. 4.22** The eye arrangement of sheet-weavers (Linyphiidae) viewed from above (**left**) and from the front (**right**). (*Figure* Miriam Frutiger)
22a	Mouthparts are dark, front body above is reddish-brown up to gray-brown, hind body is pastel-colored, yellowish, brownish, gray, or greenish, with a heart-spot as a marking on its back (Fig. 8.1). The first pair of legs is significantly longer than the rest, and the legs are lightly colored. All eyes are rather small, with the side eyes sometimes closer together (Fig. 4.23). Body length: 5–15 mm. Mostly in vegetation, occasionally hidden in clothing in the house. Females spin a sack-shaped nest in which they guard their cocoon. **Yellow sac spiders (Cheiracanthiidae)** (*Cheiracanthium*) (Chap. 8)  **Fig. 4.23** The eye arrangement of yellow sac spiders (Cheiracanthiidae) viewed from above (**left**) and from the front (**right**). (*Figure* Miriam Frutiger)

| 22b | Different. The hind body is often velvety, without markings or with white chevrons. The fourth pair of legs is longer than all other pairs of legs. Cylindrical spinnerets extended backwards. Nocturnal. The front side eyes are slightly enlarged, and the rear middle eyes are usually oval and noticeably whitish (Fig. 4.24). The spiders range from 4 to 17 mm long. Cocoons are lens-shaped, white or pinkish, attached under stones, and sometimes guarded by the female.
           Ground spiders (Gnaphosidae) (*Drassodes, Herpyllus, Scotophaeus, Urozelotes*)
           (Chap. 13)

    Fig. 4.24 The eye arrangement of ground spiders (Gnaphosidae) viewed from above (**left**) and from the front (**right**). (*Figure* Miriam Frutiger) |

Part II
Systematic Part

Funnel-Web Spiders (Agelenidae)

5

Funnel-web spiders are a family of medium to large (up to 18 mm) spiders, with almost 1400 species in 100 genera worldwide. They are web-building animals, and their web form also gives them their English name. Their web consists of a **retreat** open on both sides, with its edge widening on one side to form a catching sheet, so that the whole structure resembles an open-cut **funnel** (Fig. 5.1). Often, the catching sheet is anchored to the side and above with hanging threads or threads in tension, depending on the possibilities offered by the surroundings.

The spider spends almost all its time at the entrance of its retreat, with its front legs resting on the edge of the catching sheet. Through its sensory organs, it perceives any type of **vibration** from the area of the catching sheet and can decide whether it is a potential prey, a threat, or a sexual partner. In the case of a threat, it will hide in its retreat and possibly even leave through the back exit.

If the vibrations are caused by a potential prey animal that lands on the catching sheet or accidentally runs over it, the spider senses this. Through the type of vibration, the spider recognizes what type of prey it is. It will run from its retreat very quickly and directly toward the prey, wrap it in silk, and then give it a venomous bite. The prey is then brought back into the retreat, where it is eaten. In this way, the spider does not unnecessarily expose itself on the open web surface, where it could also be preyed upon by potential enemies.

The identifying features of funnel-web spiders include their long and strong legs, which give an intensely **hairy impression**. The body is usually elongated and is characterized at the back by two of its three pairs of spinnerets, which are usually clearly visible from above. The predominant colors are various shades of brown. Funnel-web spiders have eight small eyes (Fig. 4.15).

Funnel-web spiders naturally occur under stones, in rock crevices and caves, as well as under roots and in tree holes. From here, the pathway into human settlements is not far. This step has been achieved by various species of funnel-web spiders all over the world. In addition, species that often appear in houses or storage rooms are repeatedly spread worldwide via the transport of human goods.

© Association for the Promotion of Spider Research 2024
W. Nentwig et al., *House Spiders - Worldwide*,
https://doi.org/10.1007/978-3-031-70448-2_5

Fig. 5.1 Funnel web of a
funnel-web spider
(Agelenidae). (*Figure*
Wolfgang Nentwig)

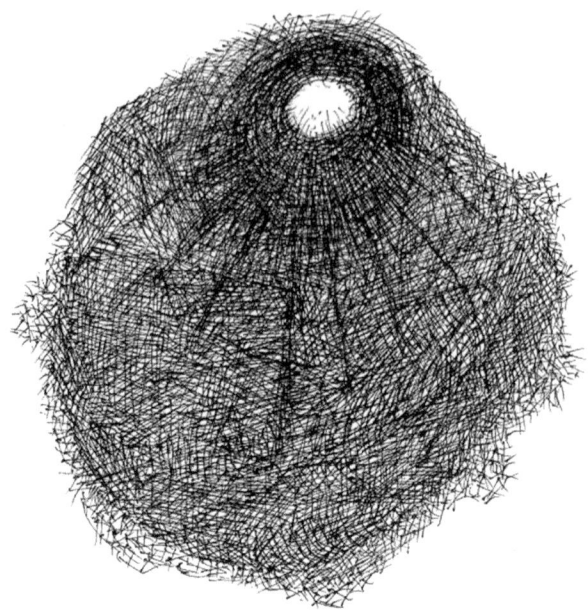

In this chapter, we deal with the North American *Agelenopsis pennsylvanica*, which exists only in North America, as well as two globally distributed species of funnel-web spiders: *Tegenaria atrica*, which is now correctly called *Eratigena atrica*, and *Tegenaria domestica*. The last two species have been very successful in colonizing human buildings in Europe and were spread to all continents via goods transport, from where they have spread further. Today, these two spiders are among the 10 most common spider species that we regularly encounter in buildings world-wide. The reasons why some *Tegenaria* species are now called *Eratigena* are discussed in more detail in Box 5.1.

Box 5.1: The *Tegenaria* Earthquake of 2013
Within the genus *Tegenaria*, an "earthquake" occurred in 2013, which strangely enough did not really interest anyone outside the small community of spider researchers, although the consequences actually affected everyone. Many of the funnel-web spiders found in buildings suddenly changed their names. What happened? Scientists had closely examined the genus *Tegenaria*, which had been known for over 200 years, along with some of their relatives, and found that an amalgamation of species had accumulated that did not actually belong together and that some species had been named at least twice. In other words, they were listed twice under different names.

In a major clean-up, formally called a **revision**, the details of which need not interest us here, the genus *Eratigena* (an anagram of *Tegenaria*) was introduced, to which about 40 species were subsequently assigned. Since then, we find in our buildings *Eratigena atrica, Eratigena duellica,* and *Eratigena agrestis* (all formerly called *Tegenaria*), as well as *Tegenaria domestica, Tegenaria parietina,* and *Tegenaria ferruginea* (all of which remained in the genus *Tegenaria*), to name just the most common of these funnel-web spiders found in houses.

Unfortunately, many of these species are difficult to distinguish from each other. Even separating the genera *Eratigena* and *Tegenaria* requires examination of the smallest details in body structure and sexual organs. Even with such large spiders, study under a binocular microscope is usually necessary. We will spare ourselves a list of the relevant characteristics here. The frequent references to color differences or differently pronounced patterns on the front and hind body of the spiders do not really help either, as these are very variable in many of the species in question. The biology of the most common species in houses, *Eratigena atrica* and *Tegenaria domestica,* is also quite similar. We therefore present both species here under their currently valid names but, of course, show photos and distribution maps of both species separately.

5.1   *Agelenopsis pennsylvanica* (Funnel-Web Spiders, Agelenidae)

Of the 14 *Agelenopsis* species, some are regularly found in or around houses, such as *Agelenopsis pennsylvanica* and *Agelenopsis potteri*. *Agelenopsis pennsylvanica* is a particularly common house spider. Both males and females are found in bushes, gardens, around the outer walls of houses, and also wander around inside houses. In cluttered basements or rooms, these funnel-web spiders like to settle on the wall behind furniture. Like many other house spiders, *Agelenopsis* often stays in rooms with high humidity, such as bathrooms, kitchens, installation rooms under houses, and basements. They always seek out the dampest spots in these places.

Agelenopsis pennsylvanica is a substantial spider species, with females reaching body lengths of between 9 and 14 mm. Males are 8 to 13 mm in size. As with many funnel-web spiders, the long spinnerets, which extend beyond the hind part of the body, are clearly visible. Males and females typically have a forebody that, in addition to light middle and side stripes, has two dark longitudinal bands (Fig. 5.2).

Agelenopsis spiders can move quite quickly, but like most spiders, they can only maintain this running speed over short distances. When an insect gets caught in their web, the spider pounces on it as it tries to free itself. As is common with web-building spiders, the male, once sexually mature, no longer maintains a web to catch prey. He patrols the surroundings in search of a sexually mature female to mate with. Occasionally, a male temporarily takes over an abandoned web.

The **venom** of *Agelenopsis* spiders is not harmful to humans. However, the spiders are large enough to penetrate our skin with their jaws. If the spider feels

Fig. 5.2 Female (left) and male (right) of *Agelenopsis pennsylvanica*. (*Photos* Sarah Rose)

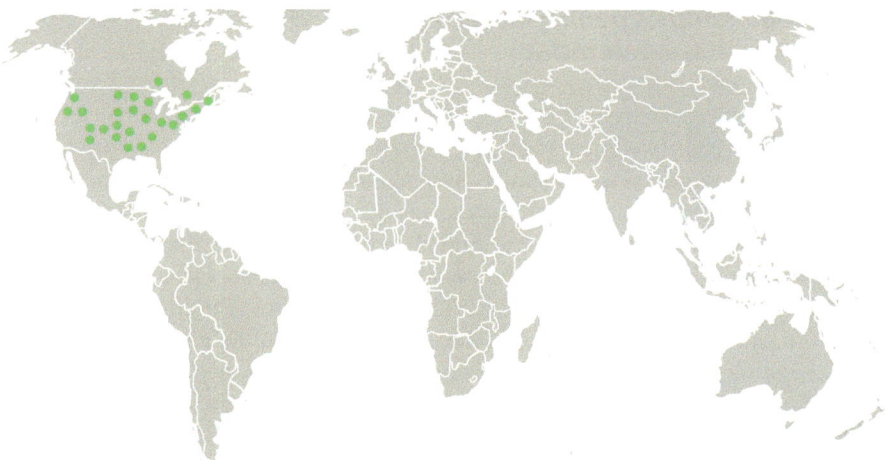

Fig. 5.3 Distribution area of *Agelenopsis pennsylvanica* (green)

threatened, bites are possible and probably somewhat painful. Further venom effects, similar to those of *Eratigena* and *Tegenaria,* will not occur.

Agelenopsis pennsylvanica is only found in the United States and neighboring areas of Canada. It is most common from Colorado in the western United States to the eastern states. Further evidence comes from the Pacific Northwest, namely the US states of Washington and Oregon (Fig. 5.3). Occasionally, reports of occurrences of *Agelenopsis pennsylvanica* outside this distribution area are reported, often only with photos as evidence. Since the color patterns of all *Agelenopsis* species are variable and very similar, photographic proof is insufficient. The only way to definitively identify the spiders of this genus is by examining their sexual organs under a microscope. *Agelenopsis* species have so far not been found outside North America. This is a surprising finding, as the similarly sized European *Eratigena* and *Tegenaria* species have been regularly displaced. Perhaps there is a peculiarity in the behavior of *Agelenopsis* species that prevents this. We simply do not know.

5.2   Giant House Spider *Eratigena atrica* and Barn Funnel Weaver *Tegenaria domestica* (Funnel-Web Spiders, Agelenidae)

In Central Europe, it is primarily the **giant house spider** *Eratigena atrica* that has conquered houses as living spaces. Within houses and gardens, you can find them in cellars, old sheds, or garages, but they are also found in upstairs rooms and under the roof. Here, the predominantly nocturnal spider builds its **funnel web** in corners, under furniture, or similar places. The catching sheet can become quite large, as it is regularly expanded and covered with a new layer of silk when the old one becomes too dusty. A clean surface of the catching sheet is important for the spider, so that the finest vibrations triggered by prey animals can be transmitted to the spider at the edge of its retreat.

Eratigena atrica, formerly called *Tegenaria atrica*, is one of the largest native spiders in Europe. For females, a body length of 11–16 mm is possible; for males, 10–14 mm. The legs can be up to 35 mm long, so a leg span, if you want to calculate it that way, of about 85 mm can result. However, the body of the male animals is slender, and both sexes appear rather gracile with their long legs.

The basic color of the spiders is dark brown in both sexes. On the front body, the animals have a light central stripe and two lateral, light stripes. The dark area in between can appear as blurry spots or be divided by light radial lines, resulting in a star-shaped pattern. The legs are medium brown or reddish-brown to dark-colored. On the hind body, some light angular spots along the back line are noticeable, which sometimes also appear as paired rows of dots (Fig. 5.4).

In common parlance, these spiders are also occasionally called "sink spider" or "bathtub spider." The reason for this is that during autumn they are often found in bathtubs or **sinks**. Spiders with such a large body size and considerable leg span always alarm people and occasionally also cause fearful reactions. Typical questions are: Why are they here? Why can't they get out of the sink themselves? How do you get rid of them?

It is noticeable that the house spiders in sinks and bathtubs are almost always mature males. You can easily recognize this by the thickened tips of the pedipalps. During autumn, spiders do not generally invade houses to find a place to overwinter, but at this time of year, it is simply the fact that the males are mature and are very actively "courting" females. Contrary to what some people fear, domestic house

Fig. 5.4 Female (below) and male (above) of the giant house spider *Eratigena atrica*. Males are easily recognizable by the slimmer rear body, the longer legs, and the pedipalps. (*Photos* above Jutta Ansorg and below Ewa Ansorg)

spiders do not come into our house through the drain from the sewer (see Box 2.3). This is occasionally suspected, but it is in fact nonsense, since these animals cannot get through the siphon, and a sewer is not a natural habitat for spiders.

It is more likely that the spiders ended up in an area of the house where they can find water. Because many spiders can generally survive a long time without food but not without water, they almost automatically end up in the sink or bathtub. If you catch these animals and release them outside, you can also give them a few drops of water and do something good.

If nocturnal domestic house spiders do somehow end up in the sink during the course of the night, then they can no longer escape on their own. They are trapped. As web-building spiders, they have claws but do not have adhesive hairs like jumping spiders and cannot climb up the smooth walls of bathtubs and sinks with only their claws. You can easily catch the spider with a glass and release it in the basement or otherwise a short distance from the house.

The **barn funnel weaver**, *Tegenaria domestica*, looks very similar to its larger cousin, and the body size is comparable in many spiders. The animals are somewhat lighter brown, often also reddish-brown or gray. The legs are occasionally darkly ringed (Fig. 5.5). The hind body is similarly marked as in the large domestic house spider. Females reach body sizes of 8–12 mm, and males 6–9 mm. Based on color or pattern alone, a distinction between the two species is not possible, as these can vary greatly.

Eratigena atrica and *Tegenaria domestica* originally come from Europe and its adjacent regions. *Eratigena atrica* has also been introduced into North America. It is conceivable that it has reached other countries too, but this is not yet known (Fig. 5.6). Due to the strong similarities between closely related *Eratigena* and *Tegenaria* species, occasional confusions cannot be excluded. Thus, the true distribution areas of many species are still inadequately known.

The situation is similar for *Tegenaria domestica*. Originally, this species of Old World origin was restricted to Europe and adjacent countries but was then spread by human transport to all continents and is now found worldwide (Fig. 5.7). One might be inclined to emphasize that the barn funnel weaver is now found in every country, but in fact, there are still considerable research gaps, especially in African and Asian countries.

Fig. 5.5 Female (left) and male (right) of the barn funnel weaver *Tegenaria domestica*. (*Photos* Gilles Arbour/natureweb.com)

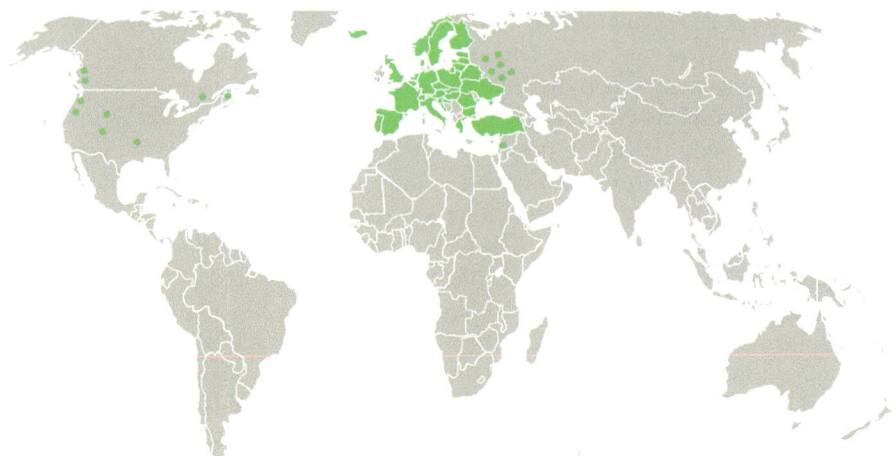

Fig. 5.6   Distribution area of the giant house spider *Eratigena atrica* (green)

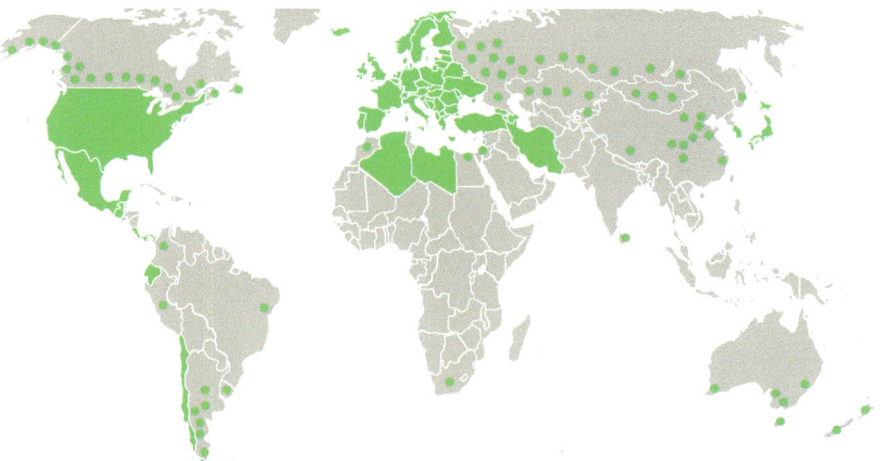

Fig. 5.7   Distribution area of the barn funnel weaver *Tegenaria domestica* (green)

The giant house spider, as well as closely related species of similar size, have strong jaws that can bite humans. Since the normal behavior of these spiders toward humans is always to **escape**, a bite will only occur if you try to catch such a spider by hand or try to remove it and accidentally squeeze the animal. But don't worry; the bite is not very painful, and occasional swelling is limited to the bite site and disappears within a few hours. Further effects of a bite by *Eratigena*—or *Tegenaria*—species are not known. The venom of all these funnel-web spider species is therefore harmless to humans (see also Box 5.2).

Box 5.2: Spider Bites: Fantasy and Reality

In the United States, the introduced field spider *Eratigena agrestis* caused quite a stir when reports of bites with considerable medical complications suddenly began to arise. This is interesting because this spider, which is otherwise very similar to *Eratigena atrica*, is widespread in Europe, but here it is rarely found in human buildings. Around 1900, *Eratigena agrestis* was unintentionally introduced to North America and spread there so quickly that it became known as the *hobo spider*, i.e., a tramp or vagrant. It was also reported that the European species behaved differently in America: being preferably found in houses, remarkably aggressive, biting easily, and with **venom** that causes nasty, slow-healing wounds. From a scientific point of view, this seemed to be an exciting story, albeit one that quickly collapsed like a house of cards. *Eratigena agrestis* is more often found in buildings in North America than in Europe. In the Old World, however, it has also been found in houses since 1950. However, their behavior has not changed; the animals are not aggressive and they do not tend to bite. Regarding their venom, there were no differences between European individuals of this species and individuals from North America. Thus, the reports of their dangerous nature turned out to be poor journalism. Not a single case of a bite with medical complications could be demonstrated with any confidence; thus, it was all just a lot of hot air.

Hackled Mesh-Weavers (Amaurobiidae)

6

Hackled mesh-weavers, Amaurobiidae, are a small spider family of just under 300 species. These rather small to medium-sized spiders are predominantly dark in color and prefer moist habitats. Hackled mesh-weavers belong to the **cribellate spiders**, meaning that they produce ultrafine crimped threads (Chap. 1). These cribellate threads give the web a bluish tinge in fresh condition, older webs appear whitish (Fig. 6.2).

6.1 Cellar Mesh-Weaver *Amaurobius ferox* (Hackled Mesh-Weavers, Amaurobiidae)

The cellar mesh-weaver *Amaurobius ferox* lives quite well-hidden within our cellars. It builds its retreat in narrow cracks and crevices of cellar walls or on the exterior walls of our houses, sometimes also under stones, beams, or among deposited rubbish. In addition, these animals can also be found in caves, mineshafts, and similar habitats. While the nocturnal spiders often live very hidden from us, their fine cribellate spider threads are quite noticeable due to their bluish tinge.

If, with a little patience, you manage to see *Amaurobius ferox* itself, the animal appears completely black at first glance. Only on closer inspection do you see the shiny reddish-brown coloration of the front body, reminiscent of chestnuts, and the black eye area with the rather small eyes (Fig. 4.17). The legs are brown and can be more reddish or more gray in some individuals. The hind body is gray-haired, which gives the animal an almost **velvety appearance**. On the back, you can see a light brown pattern of markings, consisting of paired spots and interrupted transverse stripes, sometimes also described as a heart mark or chevrons (Fig. 6.1).

With a size of 11–16 mm in females and 8–10 mm in males, which are easily recognizable by their thick, long pedipalps, the cellar mesh-weaver looks quite impressive. Once again, the saying applies: *nomen est omen. Amaurobius* comes from Greek and means "dark," while *ferox*, translated from Latin, stands for "wild, warlike." However, it's not the comparatively large human who should be afraid, but

W. Nentwig et al., *House Spiders - Worldwide*,
https://doi.org/10.1007/978-3-031-70448-2_6

Fig. 6.1 Female (left) and male (right) of the cellar mesh-weaver *Amaurobius ferox*. (Photos *Gabor* Kovács)

the small cellar woodlice, which are among their preferred prey. Depending on what's on offer, all sorts of insects, including firebugs or even wasps, are hunted.

The web of the cellar mesh-weaver consists of a wide-meshed **funnel web**, in the center of which lies the tube, open at both ends, which serves as a retreat (Fig. 6.2). From this retreat, the spider waits for passing prey that become tangled in the extremely fine cribellate threads. As soon as the spider detects vibrations from a prey item that has become entangled, it rushes out of its retreat onto the prey and tries to overpower it. This reaction can also be provoked by touching the web with a tuning fork, thus creating vibrations that resemble those of a prey animal. However, the clever spiders usually quickly learn that a tuning fork is not prey and only fall for this observation trick once.

If danger approaches, the cellar mesh-weaver tries to escape into its retreat as quickly as possible. If this is not possible, it simply drops and lands, for example, at the foot of the wall on the ground. There, the spider pulls in its legs and plays dead. However, if it is grabbed or squeezed so that it feels seriously threatened, *Amaurobius ferox* switches from flight to fight and defends itself by biting, including any human who tries to grab it with their fingers. Such very rare bites can feel like a wasp sting but are usually completely harmless. Apart from redness or swelling, the pain subsides after a few hours.

Fig. 6.2 Older and slightly dirty capture web of the cellar mesh-weaver *Amaurobius ferox*. In the area of the retreat, some fresh, blue-tinged threads are visible. (*Photo* Gabor Kovács)

In spring, the adult male sets out in search of a mating-ready female. If the search is successful, he begins to drum and twitch excitedly with his legs on the edge of the funnel web in front of the female's retreat until the female is lured out by the whole performance. It can take several hours until the female, in an ideal case, decides to accept this suitor. She is now ready to mate, allows the male to come closer, and the mating itself takes place relatively quickly, lasting only a few minutes. Afterwards, the male quickly retreats to avoid being eaten as dessert. In captivity, males are regularly eaten by the female, but under natural conditions, most seem to escape in time.

The brood care of the cellar mesh-weaver is quite remarkable. The female lays the eggs in a whitish **cocoon**, which she packs and protects well with spider threads. The cocoon is attached and guarded within the retreat. After about 3 weeks, the young spiders hatch. The female then lays a second egg ball as "first food" for her offspring, which is immediately eaten by the young. They molt 3–4 days later. Afterwards, the mother sends out various plucking signals and prepares a "bed" of spider silk for her young. They react with intense movements to the plucking signals of the mother. On the first or, at the latest, on the second day after their molt, they then eat their own mother! About an hour before her death, the mother begins with intense knocking, shaking, and jumping movements, which increase in her last 20 minutes of life. The young then gather into a dense swarm, against which the mother presses her hind body. The young climb onto their mother, who is now

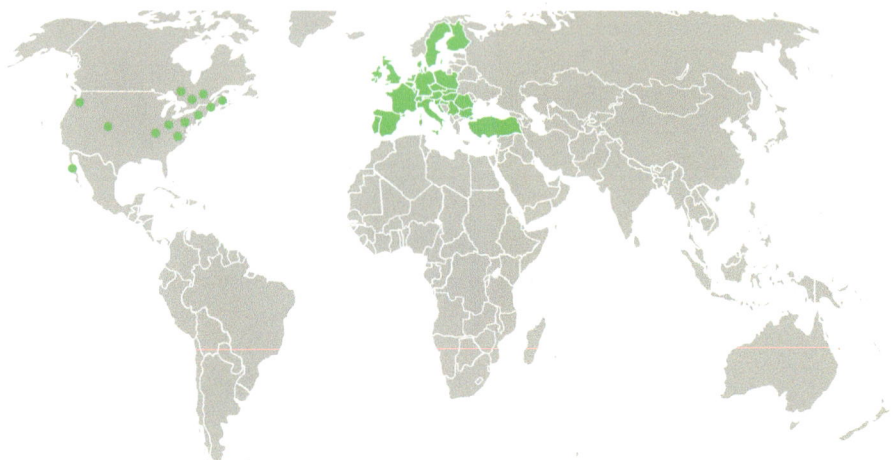

Fig. 6.3 Distribution area of the cellar mesh-weaver *Amaurobius ferox* (green)

completely relaxed. Then, as a final act, they begin to suck the juices out of their motionless mother, a process that lasts about an hour.

The body weight of the young spiders increases almost threefold after ingesting their mother! Following their cannibalistic meal, the young animals stay in the maternal nest for another 3–4 weeks and use it to overpower and eat prey animals together. Only after two more molts do they begin their life as loners.

Interestingly, the mother spider only dies if she is left with her offspring. If the mother is separated from her young after the first molt (a short time before she would normally be eaten), the mother not only stays alive but also builds another egg cocoon, from which more young spiders hatch. But why does the mother stay with her young and sacrifice herself? Would it not be better to bring more children into the world? Experimental studies have provided an answer: The overall survival rate of her offspring and their future hunting success are significantly better if the female only builds one egg cocoon and allows herself to be eaten by her young. It is indeed a maternal sacrifice with deadly consequences!

Originally, *Amaurobius ferox* was found under stones in forest areas or in caves but then quickly colonized walls and basement rooms as an excellent substitute habitat. It is widely distributed in Europe and was probably already transported to Canada, the United States, and Mexico in the early nineteenth century or even earlier (Fig. 6.3).

In addition to *Amaurobius ferox*, in Europe you can also find the closely related species *Amaurobius fenestralis* and *Amaurobius similis* in houses, as well as other species in North America. Furthermore, there are many other *Amaurobius* species found outside of buildings. Their identification is only possible through studying their genital morphology.

Orb-Weavers (Araneidae)

7

The orb-weaver spider family is the third most species-rich in the world and includes over 3100 species. This includes exceptionally colorful, pretty spiders and, in addition to small to medium-sized animals, also some quite large species. Orb-weavers are found throughout the world in almost all habitats. A number of species have also become adapted to living outside on human buildings, although they are rarely found inside houses.

Among the orb-weavers, there are a number of peculiarities, but above all the eponymous **orb web** (Fig. 7.1), in which a minimum of material is used to achieve a maximum of capture efficiency. After exploring a suitable spot, the orb-weaver draws out a horizontal thread for the construction of a new web, for example, between two branches. In the middle of this thread, it attaches another thread, abseils down on this, and pulls the whole thing downwards, so that a star-shaped structure similar to the Mercedes car logo is created – the basic framework of the orb web. The center of the "Mercedes star" later becomes the central hub of the web. Next, more radial threads and the frame threads are built. After constructing a wide-meshed auxiliary spiral (from inside to outside), in a final step, the auxiliary spiral is gradually removed (from outside to inside) and replaced by a close-meshed sticky spiral. The sticky droplets, which we then observe at regular intervals on the spiral thread, make orb webs perfect insect traps.

Most orb webs are oriented vertically. The spider can, depending on the species, be active during the day or night. It then builds a new web each day, either in the morning or during the evening twilight. Web construction takes less than an hour, and the spider then usually sits in the middle of the web, on the **hub**, waiting for prey. Some species have moved their waiting place to the side into a better-protected retreat and connect themselves to the hub via a signal thread, which is the place where all the vibration signals converge.

Large prey is individually overwhelmed by the orb-weaver, wrapped up, paralyzed or killed by an injection of venom, and then digested. Small prey animals, which can no longer free themselves from the adhesive glue, are initially ignored. But when the hunting time is over (in the evening or the next morning), the remains

© Association for the Promotion of Spider Research 2024
W. Nentwig et al., *House Spiders - Worldwide*,
https://doi.org/10.1007/978-3-031-70448-2_7

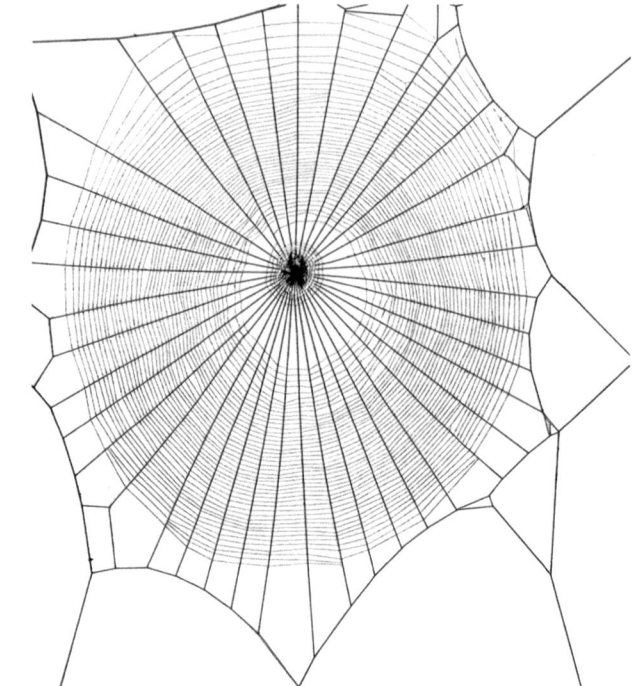

Fig. 7.1 Orb web of a typical orb-weaver spider. (*Photo* Samuel Zschokke)

of the usually damaged orb web, along with all adhering small insects, are wrapped up and eaten. Since a new orb web is built the next morning or evening, this results in a perfect recycling of the amino acids needed to build an orb web.

Orb-weavers have developed seven different types of silk glands. By combining the different types of silk, they can, depending on their needs, **produce silks** with very different properties: adhesive threads for the spiral of the orb web, particularly strong threads for the frame of the web, attachment points for securing web construction, wide silk bands to bind prey, different stable threads for various layers when building the egg cocoon, and also for safety threads for when they allow themselves to fall from the web in case of danger, and by which they can subsequently climb back up on these safety threads again.

Some orb-weavers have a robust hind body, often with hump-like protrusions on the back or on the sides. The legs of orb-weavers are rather short and often equipped with quite strong spines. This allows the spider optimal movement in its orb web and in the vegetation but also facilitates a safe approach when overpowering prey. Such legs are less suitable for running on the ground, where orb-weavers then appear rather clumsy.

7.1 European Garden Spider *Araneus diadematus* (Orb-Weavers, Araneidae)

This species appears with extremely diverse coloration, with the typical white cross on the upper side of the hind body almost always visible. Otherwise, its **body color** can vary from almost white, through yellow, greenish, orange, rust red, brown, to almost black. Black animals are found mainly in high mountains, where their color protects them well from strong UV radiation. Once, a spider expert from the Natural History Museum Bern (Switzerland) was even shown a cherry-red specimen by an outraged visitor. She had previously sent a photo of the spider to the expert by email, which he commented upon with the words "Photoshop makes it possible..."

The forebody is usually light to medium brown or gray and densely haired. The light legs are brown to dark brown, annulated, and have many short spines (Fig. 7.2). Just as variable as the color is the body size of European garden spiders. Females have body lengths of 20–23 mm, making them quite large; males, however, only reach 4–11 mm.

The European garden spider inhabits all of Europe and the temperate zones of Asia through to the Far East, including Japan. It is also not rare near the Atlantic and Pacific coasts of North America, where it was, however, introduced about 150–200 years ago, probably with sailing ships in whose rigging the spiders lived (Fig. 7.3). The spider is very adaptable and is therefore found in a variety of

Fig. 7.2 Female (**left**) and male (**right**) of the European garden spider *Araneus diadematus*. (*Photos* left Dragiša Savić and right Hans-Ulrich Kohler)

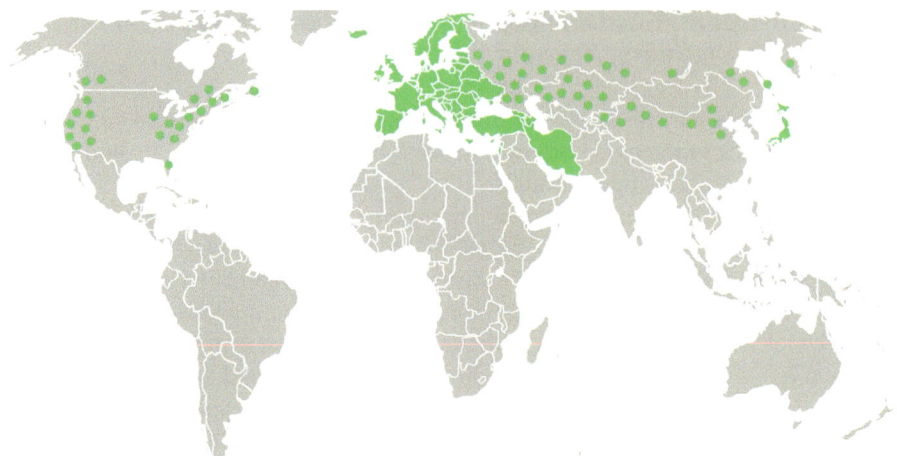

Fig. 7.3 Distribution area of the European garden spider *Araneus diadematus* (green)

habitats. It can be seen in forests and orchards, on forest edges, hedges, and in gardens. It also likes to spin its beautiful orb web on house walls and on outbuildings. In adult females, the upper frame can measure several meters, and the typical orb web itself can have a diameter of 50 cm. The spider usually doesn't have a retreat and sits on the hub in the center of the web.

The typical common name of the European garden spider comes from the fact that it likes to inhabit natural gardens. It is also sometimes referred to as the cross spider due to the typical white cross, which most specimens show on the front of the hind body. This cross consists of **guanine**, a waste product, which is also contained in the spider's feces. Part of the guanine is stored in the form of tiny crystals in converted intestinal cells, which lie directly under the thin cuticle forming the exoskeleton. The white color of the crystals is therefore clearly visible. As to the purpose of this cross-shaped marking, scientists have not reached a final agreement. Possibly, it serves to encourage flower-visiting prey, acting to both deceive and attract insects.

In myths and legends, spiders do not usually fare well. They are said to be lethally poisonous; it is claimed that they would poison food they have walked over, cause stillbirths and the plague, and that they are messengers of the devil. Of course, this is all nonsense. The garden cross spider, however, has a good reputation in some regions because of the white cross. The story goes that a spider spun a silken bandage around the wounds of Jesus Christ at Golgotha. When the spider humbly withdrew after her work, the white cross appeared on her back, which she has kept to this day. Therefore, in some parts of Austria, the garden cross spider is called "Mother God's little animal."

On the other hand, the cross is occasionally interpreted as a warning sign, as the spider is said to be poisonous. This is just as silly. Despite its size, the European garden spider has comparatively small chelicerae, with which it is barely able to bite through thick human skin. If a bite does occur on particularly delicate skin areas, only a temporary redness of the puncture site and itching similar to that after a mosquito bite can be expected.

The orb web of the garden spider is a brilliant combination of a relatively thick frame and radial threads, together with thin, elastic adhesive threads, which are arranged in a spiral (Fig. 7.4). When an insect flies into the web, the kinetic energy of the insect is absorbed by the radial threads. At the same time, these transmit vibrations into the web center, the hub, where the garden spider sits. The extremely elastic adhesive threads hold the insect fast. The spider hardly sees the insect but relies mainly on its vibration sense, which perceives the vibrating pattern of the prey.

Upon close observation, you will notice that the hub in the webs of garden spiders, and most other orb-weavers except for very young spiders, is never exactly in the **center of the web** but is shifted upwards to a greater or lesser extent. One reason for this is the weight of the spider – if it is busy below the hub, its body is pulled downwards, causing the hub to lie away from the center of the web. This was demonstrated by weighting spiders with small weights – these spiders built more eccentric webs than their unburdened conspecifics.

But that's not all: since the spider **hangs upside down** in the hub, it can reach an insect stuck in the adhesive droplets below it faster than one that is stuck above it. For this, it quickly lowers itself on a thread toward the insect. Two scientists have shown that in two related species of orb-weaver, experienced hunters build more eccentric webs than inexperienced ones. This is most likely also true for the garden spider. Thus, spiders can learn from experience and adapt their capture web accordingly!

Large insects and those able to defend themselves are wrapped with a special type of spider thread, the prey-binding threads, before the spider applies its

Fig. 7.4 Orb web of the European garden spider *Araneus diadematus*. (*Photo* Jutta Ansorg)

venomous bite. Sometimes such wrapped insects are not immediately bitten but are hung on the edge of the web in a kind of pantry as a reserve for hard times. There, the unfortunate victims wait until the spider becomes hungry again.

If a **sexually mature male** finds a female, caution is required. He slowly approaches the web in which the female sits and draws out a courtship and mating thread from the edge of her web into the vegetation. He then begins to pluck at this thread with a species-specific vibration pattern. Vibrations are transmitted through the web to the female. If the female is ready to mate, she slowly approaches and finally suspends herself on the mating thread. Then there are two – rarely more – matings, after which the male immediately jumps from the thread. Better safe than sorry…

The European garden spider usually has a two-year life cycle. About 50 eggs overwinter in an artfully spun **egg cocoon**, which the female builds one or more of toward the end of her life in the autumn. In spring, the tiny young spiders hatch from the egg cocoon and stay together at the cocoon until the second molt has taken place. If disturbed, they literally scatter and then regroup at the cocoon. After the first molt, they are strikingly yellow with a dark spot on the abdomen. During this time of peaceful coexistence, they still feed on the yolk supply in their gut. After the second molt, they become "spider enemies" and begin their own lives. Now their siblings are just potential prey. They overwinter as adolescents and develop into sexually mature animals the following year.

7.2   *Argiope* Species (Orb-Weavers, Araneidae)

Currently, 88 *Argiope* species are known, which occur on all continents, especially in warm and open habitats. Many species in this genus are strikingly colorful and very beautiful. In addition, females stand out due to their size, while males are so much smaller that they are referred to as dwarf males.

In addition to the colorful appearance of *Argiope* species, their orb webs are noticeable because they contain additional silk elements that are rarely found in other orb webs. This is a strong, zigzagging silk thread spun onto the orb web, which in some species goes from top to bottom, with or without interruption in the area of the hub through the orb web. In other species, two threads are laid in an X-shape through the orb web (sometimes only one to three separate arms are produced), while in other species there is a circular silk pattern, which is referred to as a shield (Fig. 7.5).

If a spider has built an X-shaped silk structure consisting of four individual arms, it may be that the same spider builds a structure of only one, two, or three arms the next day, or perhaps no such additional silk structure at all. Also, individuals of the same species in the same habitat differ as to whether and how they apply this additional silk thread. In summary, these structures are referred to as the **stabilimentum**, which implies that they stabilize the orb web. Since almost all other orb-weaving spiders manage without such web reinforcement, this explanation is doubtful.

It is therefore interesting that a given *Argiope* species can build different stabilimenta throughout their lives, and these may vary depending on the

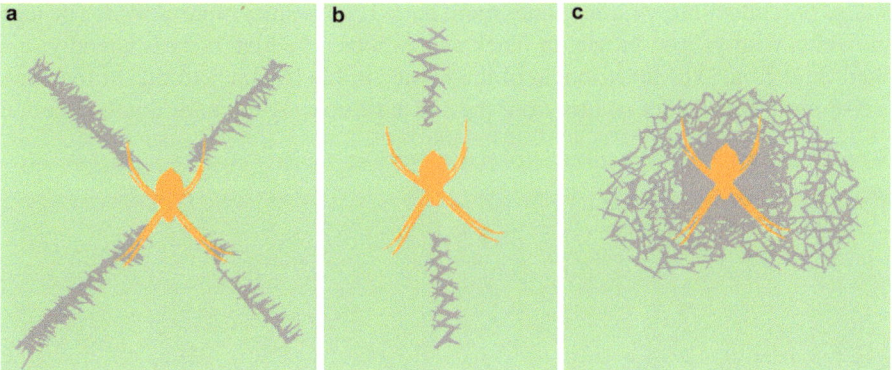

Fig. 7.5 The stabilimentum in the orb web of *Argiope* species can be (**a**) X-shaped (with one to four arms), (**b**) vertically linear (with one to two arms), and (**c**) shield-shaped. Schematic representation without capture spiral. (*Figure* Wolfgang Nentwig)

environment, the degree to which they are satiated after eating, or their molting state. It is also interesting that a single spider sometimes does not build a stabilimentum for weeks. Over the last 150 years, more than a dozen **different hypotheses** have been presented on the function of these stabilimenta, but none of them explains it satisfactorily. We must be honest and admit that today we still do not know what function these very conspicuous structures have. It is most likely that stabilimenta help the spider to camouflage itself (against certain visually oriented enemies? at certain wavelengths of light?) or that certain prey animals are attracted to these silken structures. After all, we know today that the silk of the stabilimenta reflects in the ultraviolet range and thus perhaps imitates certain flower structures.

Argiope species sit in the middle of their orb web on the hub and wait for prey. They hold their legs in pairs at the front and back together, so that they form an X. If the spider has also built an X-shaped stabilimentum, the spider merges into it. If danger approaches, the spider can, depending on the degree of threat, rapidly switch to the other side of the orb web, or it can make the orb web vibrate by intensive forward and backward movements such that the web and spider become blurry for the enemy. It can also run up on the frame threads or drop to the ground.

Argiope species catch a wide range of prey animals. Often their webs are found in meadows, hedges, and waysides and are rarely built higher than one meter above the ground. As a result, they preferentially catch grasshoppers and flower-visiting insects, which can be skillfully overpowered by the spider. However, many other insect species, often very small ones, also get caught in the orb webs.

Argiope species are not really house spiders, as they do not occur inside buildings. Nevertheless, they are regularly found on the terrace or veranda, in the conservatory, and also outside in front of the window. Since this can be observed on all continents with different *Argiope* species, we present here some species that are known to occur outside of houses.

Argiope argentata: The front body and the front part of the hind body of the female are silvery-white haired, while the rest of the rear body is gold-yellow to

orange or reddish-brown, sometimes more dotted, sometimes more striped. The legs
are white, with yellow, brown, or black annulations. The hind body is laterally ser-
rated (Fig. 7.6a). The body length of the female is 12–16 mm, and that of the male
is 5–8 mm. This spider occurs from the United States to southern South America

Fig. 7.6 Females of (**a**) *Argiope argentata* with prey, (**b**) *Argiope aurantia*, (**c**) the wasp spider
Argiope bruennichi, and (**d**) *Argiope keyserlingi*. *Photos* (**a**) Hartmut Wisch, (**b**) Mark A. Brogie,
(**c**) Michael Hohner, (**d**) Volker Framenau

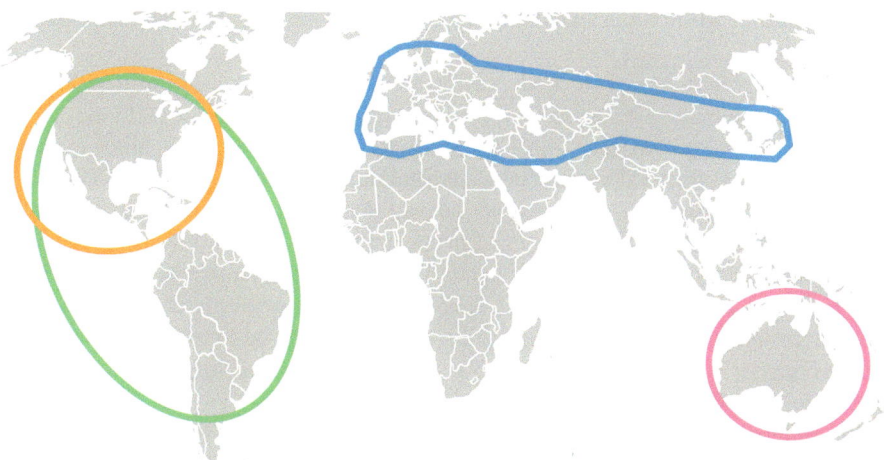

Fig. 7.7 Distribution areas of *Argiope argentata* (green), *Argiope aurantia* (yellow-brown), the wasp spider *Argiope bruennichi* (blue), and *Argiope keyserlingi* (pink)

(Fig. 7.7). It builds an X-shaped stabilimentum, usually consisting of four separate arms.

Argiope aurantia: The front body of the female is silvery-white haired, and the hind body has a stripe and dot-shaped yellow-black pattern. The legs are dark, with light stripes or annulations. The hind body is elongate-oval (Fig. 7.6b). The body length of the female is 19–28 mm, and that of the male is 5–8 mm. This spider occurs in North and Central America, from Canada to Costa Rica (Fig. 7.7). The stabilimentum consists of two vertical stripes, one above and one below the hub, which can also be connected in the area of the hub. Young animals also build a shield-shaped stabilimentum.

Wasp Spider *Argiope bruennichi*: The female's front body is silvery-white haired, and the hind body is white or yellowish with black transverse stripes, such that the spider superficially resembles a wasp. The hind body is elongate-oval, and the legs are yellowish with black annulations (Fig. 7.6c). The body length of the female is 10–25 mm, and that of the male is 7–8 mm. In Europe, the wasp spider has shown clear northerly expansion tendencies for several decades, which is probably due to human-induced climate change with warmer summer months and significantly milder winters. Today, it is found in almost all of Europe, the Caucasus, Iran, Central Asia, China, Korea, and Japan (Fig. 7.7). The stabilimentum of the wasp spider consists of a shield in young animals and, in older animals, of two vertical stripes, one above and one below the hub.

Argiope keyserlingi: The front body is covered with fine silvery-gray hairs on the female, and the rear body is white or yellowish with brown (sometimes reddish-brown or even black) transverse stripes. The rear body appears pentagonal when viewed from above. The legs are yellowish-light brown with dark brown annulations (Fig. 7.6d). The body length of the female is 10–16 mm, and that of the male is 3–4 mm. It is found in Australia and is especially common in park-like garden landscapes in urban areas (Fig. 7.7). Young animals build a shield-shaped stabilimentum, while older ones build an X-shaped one, usually consisting of four separate parts.

Argiope lobata: The front body of the female is silvery, and the rear body is white, sometimes showing a yellowish to brownish contrast coloration in the area of the lobes (lateral protrusions, which are connected in a bulging manner over the back). The legs are yellowish-brown and annulated black (Fig. 7.8a). The body length of the female is 17–26 mm, and that of the male is 5–8 mm. This species has a vast distribution area from southern Europe and Africa to Central Asia, India, and China. It probably does not occur further east (Indonesia, Australia). The distribution in Africa is currently completely unclear, so the known distribution areas are not yet connected (Fig. 7.9). The spiders usually build a simple stabilimentum, which consists of one or two silk stripes above or below the hub. However, one or two additional stripes can be added.

Argiope trifasciata: In the female, the anterior and posterior body appear silvery, and the hind body, whose sides are sometimes wavy, can additionally have dark spots, black lines, and yellow stripes or spots. The legs are darkly annulated (Fig. 7.8b). The body length of the female is 15–24 mm, and that of the male is 4–5 mm. The original distribution area of this species was North, Central, and South

Fig. 7.8 Females of *Argiope lobata* (**left**) and *Argiope trifasciata* (**right**). (*Photos* left Dragiša Savić, right Gilles Arbour/natureweb.com)

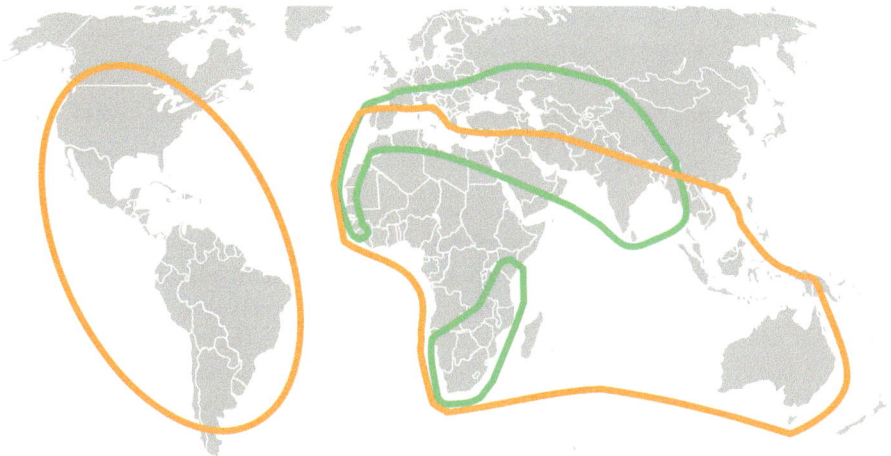

Fig. 7.9 Distribution areas of *Argiope lobata* (green) and *Argiope trifasciata* (yellow-brown)

America. However, through human activities, it was probably repeatedly brought to Africa, to Southern Europe and the Middle East, eventually to Asia (especially to India and Indonesia), and finally also to Australia (Fig. 7.9). Thus, *Argiope trifasciata* is possibly the *Argiope* species with the largest worldwide distribution. However, for most regions of Africa and Asia, it is not known whether this species occurs there. *Argiope trifasciata* usually builds a stabilimentum, which consists of one or two silk stripes above or below the hub.

The very small **males** of these six species are 3–8 mm in size, have relatively long legs for their body size, and are not conspicuously colored. Since adult males no longer build webs and are only in search of females, they (almost) no longer consume food. Their abdomen is, therefore, very narrow. You can occasionally see the males in copulation with a female or at the periphery of a web of an adult female. It is possible that the male wants to defend "his" web or "his" female against competition or is just on the way to the center of the web and thus to the web owner.

Argiope females are known as mate killers. In *Argiope bruennichi*, for example, the male never survives its first mating. As soon as he inserts his palp into the female's genital opening, he is wrapped up by the female and usually sucked dry afterward. In about a third of cases, the male manages to escape at first, usually suffering the loss of one or even several legs. The female then immediately resumes the mating position. The "love-crazy" male approaches a second time or seeks out another female. However, it never survives two mating attempts, as it is eaten either by the first or the second female. At least he can, whilst dying, still pump enough sperm into the female's reproductive organs.

7.3   Bridge Orb-Weaver *Larinioides sclopetarius*
(Orb-Weavers, Araneidae)

If you see numerous orb webs close together on bridges, fences, and walls near or directly over bodies of water, it is usually the **bridge orb-weaver**. If there are enough prey animals available, as is often the case on illuminated bridges over a river, the frame threads of neighboring webs are sometimes even used communally. With particularly closely built webs, the shape sometimes suffers, and instead of round orb webs, you can find all kinds of variants of squashed, compressed, and compact webs; sometimes, they are even reduced to relatively chaotic remnants.

In such **colonies**, where up to a hundred animals can live, the prey is never caught together, but each spider hunts for itself. So it is "only" **parasocial behavior**. Such behavior is more common in tropical and subtropical spider species and less common in species of the temperate zone. It resembles social behavior because of the dense cohabitation of the animals, but since these spiders are still aggressive toward neighbors who become too intrusive, it remains "para" social.

Bridge orb-weavers are usually active at dusk and night. During these times, they stay in the hubs of their webs, while during the day they hide outside the webs in cracks or crevices of neighboring structures, such as on bridge railings or fences.

In a bridge orb-weaver colony, animals of all ages live and thus also animals of different sizes. One might imagine that communal life offers advantages for a particularly small spider species against predators. But adult females of the bridge orb-weaver can become quite large, with a body length of 10–19 mm, so they can defend themselves quite successfully against many enemies. Males are significantly smaller than the females, with a body length of 6–8 mm.

Bridge orb-weavers have a gray to gray-brown coloration, with a striking pattern and clearly visible hair. The front body is black with white hair, which emphasizes the margins as well as a V-shaped marking toward the eyes in a stripe pattern. The hind body shows a striking foliate pattern on the rear part and an angular spot on the front part. These patterned elements are usually set off in black and clearly outlined in white. The legs can be light brown to dark brown-black and are annulated (Fig. 7.10a).

The bridge orb-weaver is distributed from Europe to the Far East in China, Korea, and Japan. It is also found in North America, but it was probably introduced from Europe. The narrow distribution area in the west of North America and its still expanding distribution suggest this (Fig. 7.11).

Despite its noticeable size and pattern, the bridge orb-weaver can easily be confused with three other species of the genus *Larinioides*. They have a similar appearance and live in a similar habitat. They build their orb webs in a comparable way, often on bushes near water or on man-made structures (bridge railings, fences, buildings) near water. However, none of these three other species occur in the parasocial associations described above. Therefore, if you find a large group of closely sitting *Larinioides* as described above, it is definitely *sclopetarius*.

The **reed orb-weaver** *Larinioides cornutus* differs from the bridge orb-weaver by the often less pronounced white pattern on the front body. On the hind body, the

Fig. 7.10 Female (**a**) of the bridge orb-weaver *Larinioides sclopetarius*, (**b**) of the reed orb-weaver *Larinioides cornutus*, (**c**) of *Larinioides ixobolus*, and (**d**) of the hedge orb-weaver *Larinioides patagiatus*. (*Photos* (**a**) Pierre Oger, (**b**) Dragiša Savić, (**c**) Gabor Kovacs, (**d**) Michael Hohner)

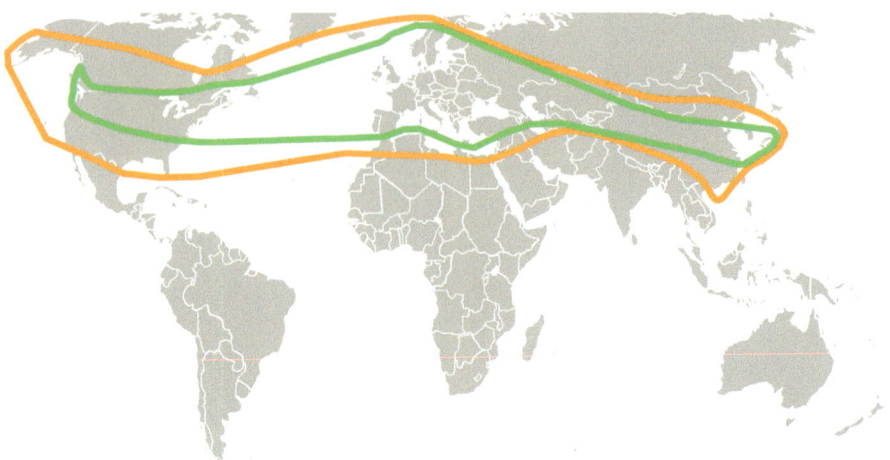

Fig. 7.11 Distribution areas of the bridge orb-weaver *Larinioides sclopetarius* (green) and the reed orb-weaver *Larinioides cornutus* (yellow-brown)

front angular spots are strongly emphasized, the dark pattern is not clearly outlined, and the color is generally very variable (Fig. 7.10b). Interestingly, the reed orb-weaver has a very large distribution area, which equally includes North America and Eurasia through to Japan (Fig. 7.11). This probably corresponds to the natural distribution area of the species.

The **hedge orb-weaver** *Larinioides patagiatus* and, as a fourth species, *Larinioides ixobolus* also show the classic foliate pattern on the hind body, but this is very variable (Fig. 7.10c, d). The white contrast pattern of the front body of *Larinioides sclopetarius* is missing in *Larinioides patagiatus*, so the front body is uniformly colored. *Larinioides ixobolus* also has a uniformly colored front body, which only occasionally shows lighter stripes. From these descriptions, it is already clear that it is unfortunately not possible to reliably distinguish *Larinioides sclope-tarius* from the three other similar species based on habitat or color. For identification, one would have to additionally examine the sexual organs.

Larinioides ixobolus is only found from Central Europe to Central Asia, and this is probably the natural distribution area of this species (Fig. 7.12). The hedge orb-weaver thus has, of all the *Larinioides* species, the largest distribution area, which seamlessly extends from North America through Europe to Asia and is only inter-rupted by the Bering Strait and the North Atlantic (see Fig. 7.12). Such a natural **circumpolar distribution** within the temperate climate zone is not often found among spiders.

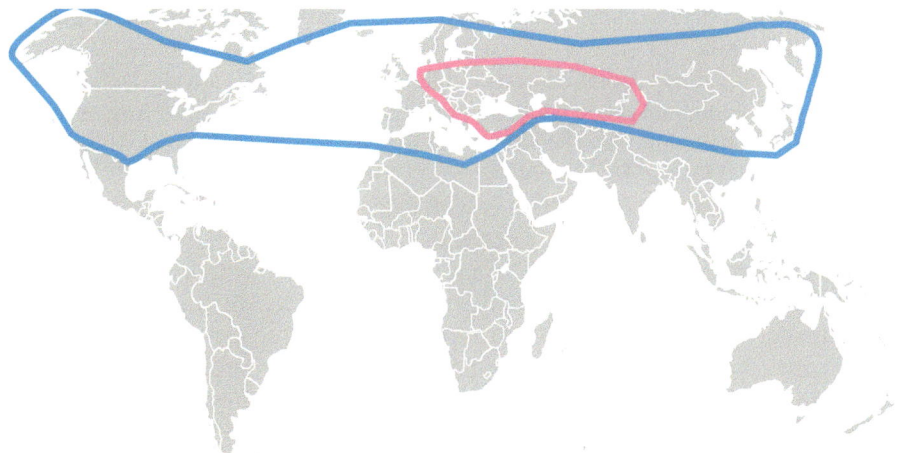

Fig. 7.12 Distribution areas of the hedge orb-weaver *Larinioides patagiatus* (blue) and *Larinioides ixobolus* (pink)

7.4   Golden Orb-Weavers of the Genera *Nephila* and *Trichonephila* (Orb-Weavers, Araneidae)

Golden orb-weavers are certainly not gold-colored, as their name suggests, but their silk is. They are not really house spiders but naturally occur in tropical and subtropical forests. They build their webs in clearings and at forest edges, as well as in lightly forested areas and transitions to bushland. Wherever there are flight paths for insects, they can find an ideal habitat for their huge webs. As human settlement areas expand everywhere and often attract insects, golden orb-weavers probably also found that there are excellent places for their webs between houses and sheds, garages, and other outbuildings. On buildings, the spiders can therefore sometimes be so common that they can hardly be overlooked.

Worldwide, there are 36 species of golden orb-weavers, which were originally all assigned to the genus *Nephila*. However, 26 species now belong to the closely related genus *Trichonephila*, while 10 species remain in *Nephila*. Regarding their biology, all species are quite similar, so we can treat them together in the same chapter. The distribution areas of some *Nephila* and *Trichonephila* species overlap. So, if you find a golden silk spider in the subtropics and tropics of this world, you should not forget that besides the four species mainly considered here, other species can occur in many areas.

In addition to their frequency (where there is one, a second spider cannot be far away), golden orb-weavers stand out due to their size and the color of their webs. Females have, depending on the species, a body length of 20–40 mm, with a noticeably elongate hind body and long, strong legs. They are thus among the largest orb-weaving spiders. Males, on the other hand, are **dwarf males** with a body length of at most 7 mm.

Nephila and *Trichonephila* species are the only orb-weavers that produce **golden yellow spider silk**, for which the whole spider group became known. It can be assumed that this is an adaptation to their habitat, as these webs are particularly well camouflaged in the half-shade between trees and shrubs. If this were true, it would be remarkable that this was only achieved by one group of forest-dwelling orb-weavers worldwide, namely the golden orb-weavers.

Golden orb-weavers build orb webs with a diameter of up to a meter. Their hub is not located in the middle, as is usual with orb webs, but at their upper edge. The spider builds the capture spiral under and next to the hub with perpendicular-like turns, so that the finished web resembles a three-quarter web rather than a classically circular orb web (Fig. 7.13). Around the web, the spider weaves additional threads, such that at a distance of more than one meter there are often barrier threads that make it difficult to approach the capture web without setting the whole thing in motion. This **barrier web** is a safety and warning system that allows the spider to escape from the hub, where it usually stays. When endangered, it flees upwards via some silk threads, for example, into a neighboring tree. As is usual with orb-weavers, the capture spiral is regularly renewed. However, since golden orb-weavers build particularly small-meshed webs, a daily renewal would be very laborious. Therefore, these spiders usually only renew one half of the web per day.

Nephila and *Trichonephila* species are well adapted to a **life in semi-shade** between branches. In *Trichonephila clavipes*, the basic color of the body is greenish olive brown, the black front body is silvery-white haired, and the hind body has

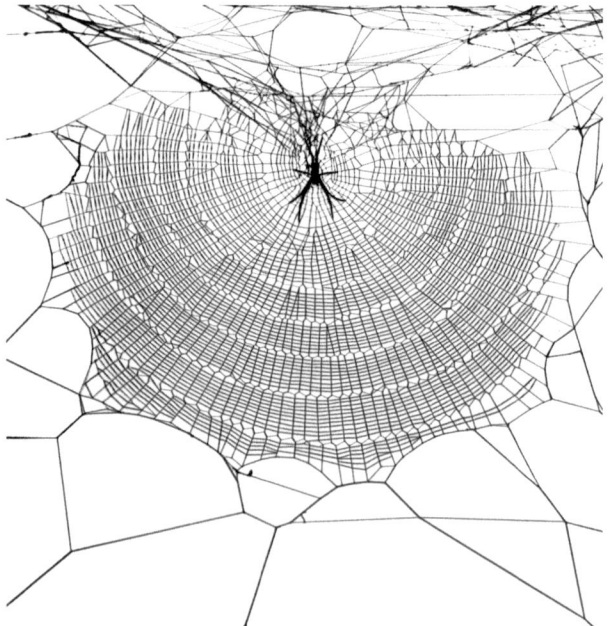

Fig. 7.13 The orb web of *Trichonephila clavipes*. (*Photo* Samuel Zschokke)

white or yellow spots. The legs are yellowish and annulated, with characteristic black hair tufts (Fig. 7.14a, b). In *Trichonephila senegalensis*, the basic color is brown to black with large, whitish to yellow spots. The forebody is covered with white hairs, and the legs are annulated yellow-black (Fig. 7.14c). *Trichonephila*

Fig. 7.14 Females of (**a**, **b**) *Trichonephila clavipes* and (**c**) *Trichonephila senegalensis*, with a male in the background. (*Photos* (**a**) Mike Deep, (**b**) Sarah Friedli, (**c**) Matjaz Kuntner

clavipes is a widely distributed species in America, while *Trichonephila senegalensis* occurs in Africa (Fig. 7.16).

In *Nephila pilipes*, the rear body is gray-brown colored, often with two to four yellow longitudinal stripes on the back and yellow dots on the belly. The front body is silvery white, the palps are orange, and the legs are black with a few yellow dots (Fig. 7.15). *Trichonephila edulis* has a white to light brown basic color, with legs that are white-brown striped and have black hair tufts (Fig. 7.15). Both species occur in Southeast Asia and Australia (Fig. 7.16).

In all the species presented here, the color pattern can vary considerably. Although the spiders are large and appear strikingly colorful to us, they are remarkably well camouflaged in their natural habitat with its rapidly changing light conditions in their semi-shaded environment.

Like all orb-weavers, the capture spiral in the web **contains glue droplets**. Small insects flying into the orb web immediately become stuck and usually cannot free themselves. The spider can take its time and only occasionally collects whatever has accumulated. This occurs at the latest during the removal of the old capture spiral and its reconstruction. These small insects are in any case beneficial because the spider eats the old capture spiral along with all the insects stuck in it.

Large insects, on the other hand, could struggle free again and require a quick reaction. The golden orb-weaver, lurking at the edge of the hub and placing its forelegs on different web sectors, feels through the **vibrations** of the prey and the

Fig. 7.15 Females of *Nephila pilipes* (**left**) and *Trichonephila edulis* (**right**). (*Photos* left Gordon Ackermann, right Ondřej Michálek)

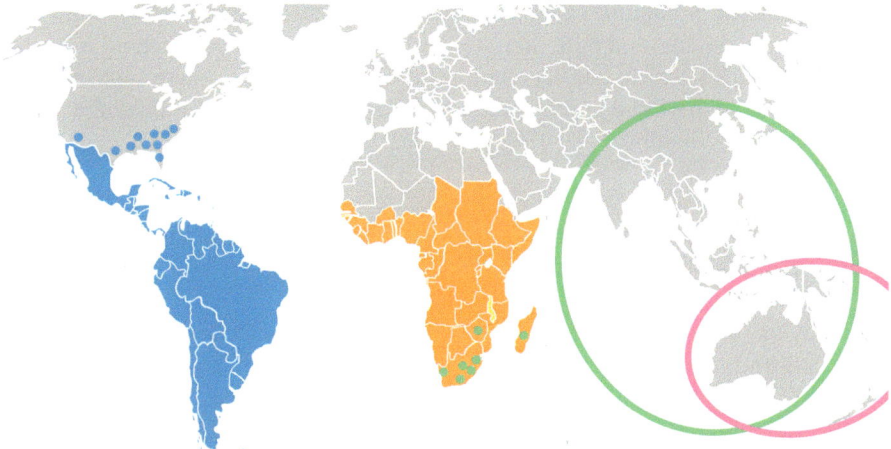

Fig. 7.16 Distribution areas of *Nephila pilipes* (green), *Trichonephila clavipes* (blue), *Trichonephila edulis* (pink), and *Trichonephila senegalensis* (yellow-brown)

different signals in the different legs, exactly where in the web the insect is wriggling. It rushes toward the prey at lightning speed. Prey that are able to defend themselves are wrapped in silk threads and thus made incapable of fighting, then bitten. Others are first bitten and later wrapped. In the end, the paralyzed or dead prey is dragged to the hub and eaten there in peace.

The orb webs of the *Nephila* and *Trichonephila* species mainly catch small insects like flies and mosquitoes, which make up about 90% of all insects caught in their webs. The remaining **prey** are significantly larger and worth catching individually: bees, wasps, beetles, butterflies, grasshoppers, and cockroaches. In very rare cases, even small birds and bats can be caught. In a one-year study on *Trichonephila clavipes*, it was found that the average prey size was 2 mm, and only 2% of the prey were over 10 mm in size.

The webs of the golden orb-weavers, already remarkable due to their color and size, are a kind of microcosm of their own. In addition to the large female, one or more **dwarf males** are regularly found in the webs. They prefer to stay near the female; sometimes a male even sits on top of the female, so to speak, to show off his possession against the competition.

Furthermore, in almost every web live several species of **thief spiders** of the genus *Argyrodes* (cobweb spiders, Theridiidae), which are usually easy to recognize by their black-silver coloring and their immobile body posture with forward-stretched legs. Some look like shiny droplets (an alternative name is dewdrop spiders), waiting at the periphery of the web for things to come. When small prey animals get caught in the capture spiral, these tiny thieves, only a few millimeters long, often approach very slowly in order not to send any vibration signals to the web owner, and suck on the prey. Even ensnared living prey animals, which the web owner has hung up for storage, are secretly sucked dry by thief spiders. *Argyrodes* species are food thieves and are therefore also referred to as

kleptoparasites. Sometimes they even sneak up to the mouth opening of the web owner to suck on the food pulp that the golden orb-weaver has prepared for itself with its own digestive juices secreted onto the prey item. So here, not only the prey but also the digestive juice is being taken; a true example of food being stolen right out of your mouth.

7.5   *Nephilingis cruentata* and *Nephilengys malabarensis* (Orb-Weavers, Araneidae)

The body of the gigantic *Nephilingis cruentata* is brown-black, with numerous white dots and spots on the hind body, which can merge into a light gray leaf-like marking. The front edge of the hind body can show a yellow line, and there are even individuals with a yellow contrast pattern on the back. The legs are partly reddish brown, orange, and sometimes also strikingly annulated (Fig. 7.17). The most

Fig. 7.17 Females of *Nephilingis cruentata* (**above**) and *Nephilengys malabarensis* (**below**). (*Photos* above Matjaz Kuntner, below Gordon Ackermann)

astonishing thing, however, is the underside, where, in addition to yellow spots on the hind body, there is a bright yellow, orange, or blood-red belly plate. *Nephilingis cruentata* reaches a body length of 28 mm, along with its long, strong legs, and is thus one of the largest spiders known. The associated males are **dwarf males** with a body length of 4 mm, who like to stay in the female's web, sometimes even directly next to or on the female.

Originally, the genus *Nephilengys* included species from tropical Africa and Asia. In 2013, however, this group was divided into an African (now four species) and an Asian branch (two species), with the African part being renamed *Nephilingis*. No, this is not a joke, but it does not make dealing with the scientific names any easier. The biology of these species is quite similar, so we can discuss them together here.

Another striking feature of these spiders is a rather large orb web, which is remarkably **asymmetric**. The hub is always at the top edge, often attached to a solid substrate with a tangle of threads leading into a tube. The spider naturally builds its web on large tree trunks, where it easily leans against the trunk. The tube then runs behind a loose piece of bark, under which the spider has set up its retreat. Some webs are also built on rocks, with the retreat then located in a crevice in the stones.

From this lifestyle, it is only a small step to human buildings. Stone or wooden houses are exactly the right habitat for *Nephilingis cruentata*, which then comfortably establishes itself under the protruding roof of the house with its tube leading into a safe retreat. The orb web is expanded downwards and covers the house wall. Due to the asymmetric position of the hub, the web of these spiders is less of a (round-woven) orb web than a three-quarter web, in which the adhesive spiral is woven in perpendicularly. Depending on the situation, this can also lead to a narrow but high web, in which the turns of the adhesive spiral are laid out in a zigzag pattern.

As with *Nephila* and *Trichonephila*, to which *Nephilengys* and *Nephilingis* species are closely related, the large web is often only partially repaired when damaged. The left half can therefore be new, while the right still shows insects, dirt particles, and web damage from the previous day, and the next day it is the other way around. You often find a large number of **kleptoparasitic** cobweb spiders (Theridiidae) of the genus *Argyrodes* in their webs as well.

Nephilingis cruentata, which is widespread in the tropics of Africa, was probably already transported to South America 200 years ago and has spread in some countries there (Fig. 7.18).

Nephilengys malabarensis is a widely distributed species in tropical Asia (Fig. 7.18), but with a body length of 15 mm, it is significantly smaller. The front body is reddish-brown, and the hind body is whitish-brown with brown spots and lateral light stripes. The orange breastplate and two or three pairs of orange spots under the rear body are striking. The legs are annulated black and yellow (Fig. 7.17).

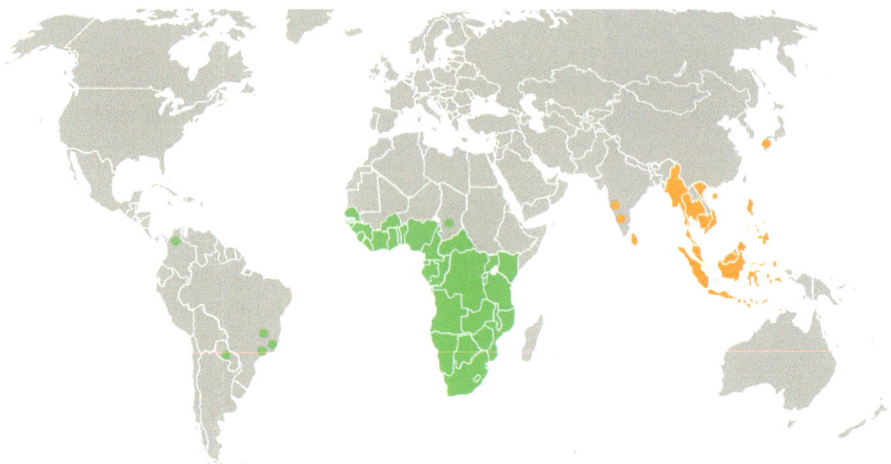

Fig. 7.18 Distribution areas of *Nephilingis cruentata* (green) and *Nephilengys malabarensis* (yellow-brown)

7.6   Walnut Orb-Weaver *Nuctenea umbratica* (Orb-Weavers, Araneidae)

Conspicuous and common everywhere in its distribution area and yet, due to its strictly nocturnal lifestyle, hardly noticed: this is the walnut orb-weaver. At first glance, it appears mainly dark, with a reddish-brown to almost black-colored hind body and dark legs, whose outer extremities are annulated reddish-brown and black (Fig. 7.19). Females grow to a size of 14–16 mm, while the males are significantly smaller at 7–9 mm but similar in color to the females. Upon closer inspection, you can see on the hind body, which has a distinctly flattened shape, a leaf-shaped pattern which is bordered by a narrow, yellowish line. In addition, there are three to four pairs of dot-shaped indentations on the hind body. Muscles attach at these points, which allow the animal to contract the hind body even further. Thus, the walnut orb-weaver manages to hide in very narrow cracks and crevices; in Germany, for example, it is commonly referred to as the crevice orb-weaver.

During the day, the spider maintains contact with its web from its hiding place via a signal thread. However, at night, it leaves its hideout and lurks in the hub of its orb web. Like the European garden spider *Araneus diadematus* described previously, it builds an **eccentric web**, whose hub is shifted upwards toward the hiding place. Webs can have a diameter of up to 70 cm. They have more spiral threads below the hub than above and have relatively few radial threads with 15–22 radii.

Fig. 7.19 Female (**left**) and male (**right**) of the walnut orb-weaver *Nuctenea umbratica*. (*Photos* Jean-Philippe Taberlet)

The hub itself is usually quite coarsely woven. Thus, such a large and coarse-meshed orb web is in many cases also visible during the day. Even if the animal is within its hiding place, it is still fairly easy to assign a web to *Nuctenea umbratica*. If you then want to see the spider, you have to follow the signal thread from the center of the hub and look for the spider's retreat.

The walnut orb-weaver lives in places where it can hide well with its flattened body during the day. These can be trees in the forest with loose bark, but also structures in human settlements, such as houses, walls, bridges, fences, and woodpiles, which have many cracks. Since the walnut orb-weaver has a two-year life cycle, it has to hibernate once, for which it chooses the same places that it also uses as a daytime hiding place. In suitable places, quite a few individuals can live in a small space, such as on house facades or under facade cladding. However, they never sit as close together as the spiders in the parasocial colonies of the bridge orb-weaver *Larinioides sclopetarius*. In places densely inhabited by walnut orb-weavers, you can certainly observe many males at night looking for females. They can engage in bitter fights amongst themselves, often until the death of one opponent (Fig. 7.20).

Nuctenea umbratica is found throughout Europe. Its eastern distribution extends to the Caucasus and Iran, and to the south into North Africa (Fig. 7.21).

Fig. 7.20 Outcome of a fight between two males of the walnut orb-weaver *Nuctenea umbratica*: the winner lost a leg, the loser lost his life. (*Photo* Jutta Ansorg)

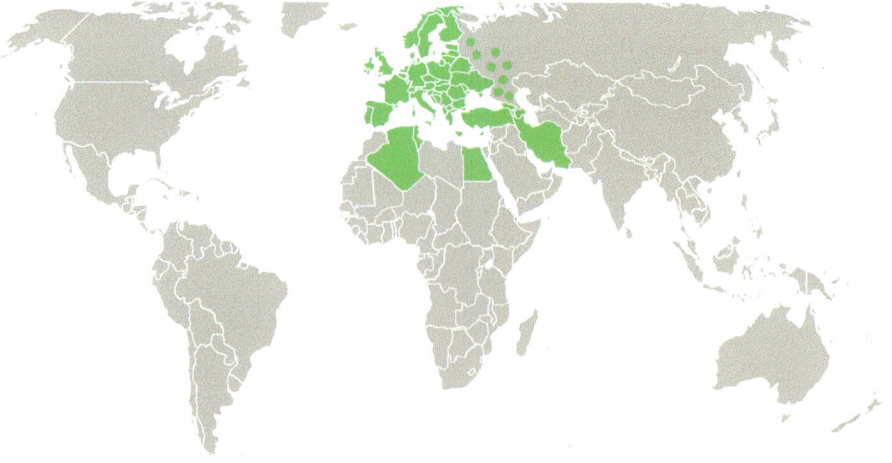

Fig. 7.21 Distribution area of the walnut orb-weaver *Nuctenea umbratica* (green)

7.7   Open Sector Orb-Weaver *Zygiella x-notata* (Orb-Weavers, Araneidae)

Often, one can see orb webs that appear to be damaged by the wind or a large prey item. However, appearances can be deceptive: if a sector is missing in the web (Fig. 7.22), usually in the upper area, it is the intentionally constructed web of an orb-weaver belonging to the genus *Zygiella*, often the open sector orb-weaver *Zygiella x-notata*. In the middle of the omitted sector, a **signal thread** leads from the center of the web to the spider's secluded retreat. Here it stays during the day, eats its prey, and also lays its cocoons. At night, it lurks in the center of its web.

It is interesting that young open sector orb-weavers still weave a complete orb web and sit in the hub even during the day. But even adult spiders weave a complete orb web, specifically when the signal thread to the retreat forms an angle greater than about 40° with the plane of the web. But why is this so? At an angle of less than 40°, there is a risk that the signal thread could become stuck to the capture spiral in strong wind, which would greatly impair signal transmission. If the angle is greater than 40°, the spider can build the complete capture thread zone.

The omitted sector is not created by the spider first weaving a complete orb web and then removing the capture spiral from a sector. Instead, it leaves it free from the outset by reversing when it reaches the radius thread of the sector to be left free when weaving the capture spiral, and then reversing again when it reaches the radius thread on the other side of the free sector. In Fig. 7.22, these reversal points are clearly visible.

Fig. 7.22 The orb web of the open sector orb-weaver *Zygiella x-notata* with a free sector and reversal points of the capture spiral. *Photo* Jutta Ansorg

The open sector orb-weaver is found almost exclusively in residential areas. Less frequently, especially in Southern Europe, it also occurs in bushes or on trees. It likes to weave its webs in corners and angles of doors, window frames, fences, balcony railings, or gutters.

With a body length in females from 7 to 11 mm and males from 4 to 7 mm, the open sector orb-weaver is not a particularly large spider species. Its appearance is rather inconspicuous. Overall, it appears silvery gray-brown. Only on closer inspection can the light-framed foliate pattern be seen on the hind body, which becomes somewhat lighter in the middle and shines with a partially silvery sheen. The front body is yellow-brown with a black longitudinal band that widens toward the front. The legs are annulated pale brown-black (Fig. 7.23).

Zygiella x-notata is very common in some places. It is found in Europe up to the Caucasus. Presumably, it was introduced early on by sailing ships and transported to the Atlantic and Pacific coasts of North America. Tropical regions of America, Africa, and Asia are apparently not suitable habitats, but in southern South America, it has become quite common in places in Chile, Uruguay, and Argentina (Fig. 7.24).

Fig. 7.23 Female (**left**) and male (**right**) of the open sector orb-weaver *Zygiella x-notata*. (*Photos* left Pierre Oger, right Hans-Ulrich Kohler)

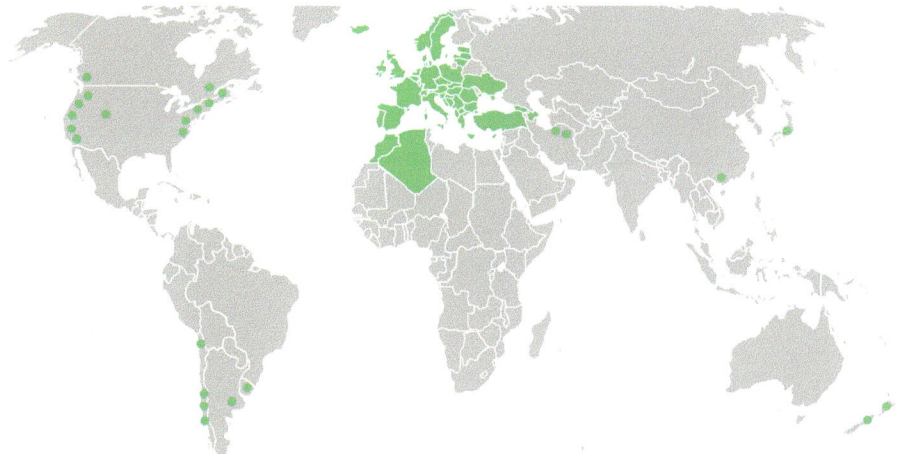

Fig. 7.24 Distribution area of the open sector orb-weaver *Zygiella x-notata* (green)

The yellow sac spider family has a turbulent history, as the classic yellow sac spiders, which are primarily discussed here, have changed their family affiliation three times in the last 30 years. Today, they form their own family, as was already proposed back in 1887. Almost 400 species are currently counted among 14 genera of Cheiracanthiidae.

Our knowledge of most of these spiders is very limited. Only a few species of the eponymous genus *Cheiracanthium* have attracted broad interest, as they occasionally appear in buildings and were quickly suspected of inflicting severe injuries on humans. Exploited through media hype, much of this denigration has stuck, and **yellow sac spiders** still have a rather poor reputation. However, meticulous scientific studies have repeatedly shown that *Cheiracanthium* species are quite harmless to humans. At least in the German-speaking world, part of the negative image of these animals may also be due to the name "Dornfinger" (thorn finger spider), which leaves much room for imagination. In reality, it refers only to a thorn-like projection on the male palp, which serves as an anchor during mating with the female spider and characterizes the genus *Cheiracanthium*.

8.1 Black-Toed Yellow Sac Spider *Cheiracanthium mildei* (Yellow Sac Spider, Cheiracanthiidae)

The northern or black-toed yellow sac spider, *Cheiracanthium mildei*, was originally a Mediterranean species and has a long history of increasing its range. In Europe, it has been expanding north of the Alps for more than 100 years and has now reached the northern parts of Germany and Poland. It was introduced to the United States more than 70 years ago and is now widespread in North America (Fig. 8.2). This species has a strong affinity for human buildings and is regularly found in our houses. In addition, some other *Cheiracanthium* species occur in our houses. Therefore, it is even possible that two *Cheiracanthium* species occur in one building in a given region. However, worldwide there are 218 different

© Association for the Promotion of Spider Research 2024
W. Nentwig et al., *House Spiders - Worldwide*,
https://doi.org/10.1007/978-3-031-70448-2_8

Cheiracanthium species. Since many of these species look similar, they can easily be confused. The only way to identify them unequivocally would be to examine their sexual organs (which we will not discuss further here).

In North American houses, we occasionally find two species: *Cheiracanthium inclusum* and *Cheiracanthium mildei*. *Cheiracanthium inclusum* is native to America, while *Cheiracanthium mildei* was introduced from Europe. They are now quite common in and around houses throughout North America. In Europe, *Cheiracanthium mildei* is widespread in houses, and *Cheiracanthium punctorium* (from Europe to Central Asia) has also been reported from settled areas, albeit very rarely. This other species has also been spreading for more than a century and is increasingly found near buildings, but not in houses themselves. It is widely believed that the two species have often been confused. In Africa, *Cheiracanthium furculatum*, a South African species, is regularly found in houses. In general, it can be said that *Cheiracanthium mildei* is becoming a common house spider throughout the world (Fig. 8.1).

Cheiracanthium mildei is referred to in English as the black-toed yellow sac spider because of its pretty, somewhat translucent gray-yellowish color. The German name "Hausdornfinger" refers to its habitat near houses. The spider has eight eyes, arranged in two rows (Fig. 4.23), darkly colored jaws, and legs with dark tips, which are occasionally compared to black socks or slippers (Fig. 8.1). Females are 5–11 mm long, while males are 4–9 mm. The black-toed yellow sac spider actively hunts at night and roams around in search of insects on vegetation or on house walls.

Fig. 8.1 Female (**left**) and male (**right**) of the black-toed yellow sac spider *Cheiracanthium mildei*. A typical identifying feature of this species is the black, slipper-like toe tips. (*Photos* left Pierre Loria and right Eckhart Derschmidt)

Since this spider actively hunts insects on leaves, it can be an important beneficial organism in controlling such insects that damage crops and flowers in orchards and gardens that have not been treated with pesticides. During the day, the spider spins a silk cocoon as a retreat in the corners of walls or ceilings in houses.

Cheiracanthium inclusum is sometimes also discovered in and around houses but is more often found in the natural environment, where the spider weaves small leaves together to build its silken hideaway. It has been observed that sac spiders drink nectar from flowers and utilize the high sugar content of this resource. The addition of nectar to their usual diet can increase the growth and survival rate of these spiders.

The closely related *Cheiracanthium punctorium* is a much larger species (females up to 15 mm body length). It has a yellow hind body and an orange forebody, which can sometimes look very conspicuous. This species lives in open heathland, pine forest edges, vineyards, and similar warm habitats, but also in overgrown areas and sometimes even in gardens. Since it rarely occurs in or on buildings, the probability of encountering *Cheiracanthium punctorium* is rather low.

The black-toed yellow sac spider is often found on the walls of houses. If someone tries to remove the spider by picking it up with their bare fingers or with a paper towel, the spider does not hesitate to bite. Yellow sac spiders have relatively long and strong **jaws**, which can easily penetrate human skin. Another typical situation occurs when you want to put on a piece of clothing in which a spider is hiding and it gets squashed in the process. The animal feels threatened and bites.

Bites from these spiders are not harmful but can be painful and cause mild swelling and itching, usually without major complications. The pain is localized and can last several minutes. One could therefore conclude that bites from *Cheiracanthium mildei* are much less painful than the sting of a mosquito. The bite of the much larger *Cheiracanthium punctorium* is, however, somewhat more painful; the local

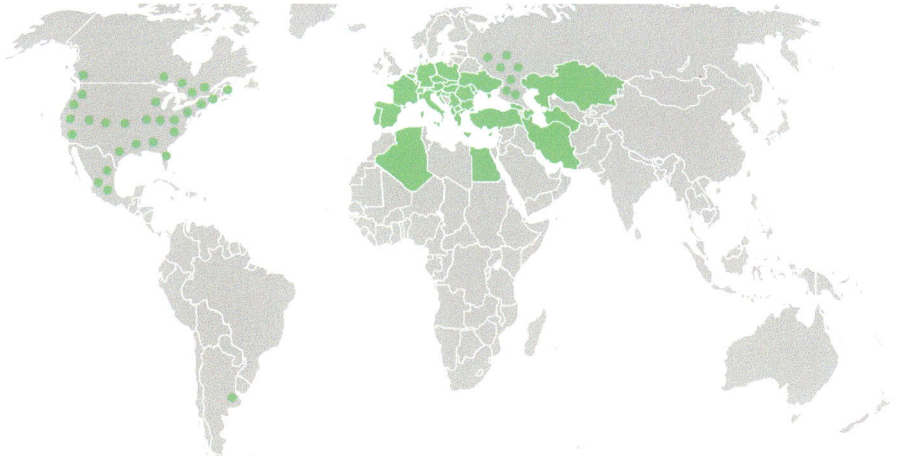

Fig. 8.2   Distribution area of the black-toed yellow sac spider *Cheiracanthium mildei* (green)

swelling can last a bit longer, and occasionally a local numbness around the bite site can radiate out a bit further. However, after a few hours, all these symptoms will have disappeared, and no long-lasting effects of the **venom** have been observed. Bites by *Cheiracanthium punctorium* are probably very rare, and corresponding reports may be due to the above-mentioned confusion with the much more common and harmless black-toed yellow sac spider.

Another aspect is, however, more serious: A few years ago, it was reported that the bite of yellow sac spiders causes severe tissue inflammation that, in turn, leads to **necrosis**, i.e., tissue death, such that legs or arms ultimately have to be amputated. However, this is real "fake news," which was often repeated in the media despite having absolutely no factual background. In other words, we have here an example of an urban legend: a completely invented story. Scientific investigations have clearly shown that the venom of *Cheiracanthium* species does not cause tissue necrosis. It is therefore appropriate to classify yellow sac spiders as harmless.

Intertidal Spiders (Desidae)

9

The species-poor family Desidae currently comprises just under 330, mostly medium-sized species and is unusual for three reasons. First, the habitat of many Desidae is in the tidal zone of the world's oceans. Spiders, as emphasized in every spider book, occur (with the exception of one species) only on land. This assessment seems to have overlooked the desids, which are very aptly called intertidal spiders. This is because they live on **mudflats**, a habitat that is only exposed at low tide. Here, they occur as free-hunting spiders. Thanks to their scopulae (adhesive hairs for good climbing on smooth surfaces), they are perfectly adapted to this habitat and catch, for example, small crustaceans. When the sea level rises again at high tide, they retreat into cavities in the ground, between rocks, or in animal structures such as empty barnacle or mussel shells. There, a silk web protects them from the water and provides the spider with an air space.

Second, the distribution of Desidae living in the intertidal zone is also unusual. They have a clear distribution focus in Australia and New Zealand, as well as tropical coastal areas in Asia and South America. It sounds like a bad joke, but since desids did not originally occur in the northern hemisphere, they were ignored here for a very long time. The fact that early and former major centers of spider research were also in the northern hemisphere is probably also the main reason why the freshwater spider *Argyroneta aquatica* was long considered as the only spider that can live underwater. Desidae use the intertidal zone of the sea, albeit differently from their freshwater counterpart.

Third, intertidal spiders are **cribellate** spiders (Chap. 1). In the tidal zone of the world's oceans, webs for prey capture are obviously not an option, and intertidal spiders live there as free-living hunters, i.e., without webs for prey capture. However, some species of desid that do not live in the tidal zone build extremely effective funnel webs with crimped threads and occur in classic terrestrial habitats. Whether this is the original desid lifestyle, or whether it is a group that shifted from the tidal zone back into land habitats, remains to be investigated.

The genus *Badumna*, with its 16 species that occur from Australia via Indonesia to China, also belongs to the land-living Desidae. At least one *Badumna* species has

© Association for the Promotion of Spider Research 2024
W. Nentwig et al., *House Spiders - Worldwide*,
https://doi.org/10.1007/978-3-031-70448-2_9

adapted to human settlement areas, and we find it in buildings on almost all continents. Therefore, we have also included it in this book.

9.1 Gray House Spider *Badumna longinqua* (Intertidal Spiders, Desidae)

At first glance, one might confuse the gray house spider, *Badumna longinqua*, with a hackled mesh-weaver, i.e., a species of the family Amaurobiidae. These approximately 10–18 mm long, darkly colored spiders (males 8–10 mm) also share another peculiarity with hackled mesh-weavers: the cribellate web equipped with ultra-thin cribellate wool (Chap. 1.3) for catching prey, which has a slightly blue shimmer. The eight eyes are small; only the front middle eyes are somewhat larger (Fig. 4.20), but good vision seems unlikely.

But first things first: *Badumna longinqua* originally occurred only in the east and south of Australia. There, it still inhabits rocks, cliffs, tree stumps, and even human settlements. It builds its web on walls, houses, roofs, fences, and in the vegetation of parks and gardens. The web can be used for a very long time, as the spider simply regularly repairs any damage to the web. And this web is something special: the spider sits in the funnel-like exit of the web and waits there for prey. From the **funnel**, one to two rather coarse sheet webs extend into the surroundings (Fig. 9.1).

Fig. 9.1 The gray house spider *Badumna longinqua*. A female (**above**) guards her eggsac; **below**, a section from the cribellate net. (*Photos* above Alvaro Laborda, below Miguel Simó)

Between the longitudinally running axis threads of this capture thread zone, the cribellum wool is zigzagged into relatively thick bundles of countless fibers of the finest silk. They are so thin that they cannot be seen with the naked eye and even under the microscope, they are not recognizable as individual threads.

Insects become almost hopelessly entangled in these ultra-fine cribellate threads. But above all, in a manner similar to a sponge, these bundles of threads absorb the semi-liquid wax layer that lies on the surface of the insect's cuticle. Thus, the web forms a firm connection with the insect's exoskeleton. The prey would therefore have to, as it were, "peel off its skin" if it were to escape from this trap.

Males of the gray house spider show a particular preference for **virgins**: even before the female has molted into an adult animal, a male enters her retreat and cohabits with her until she becomes sexually mature after the final molt. This is remarkable insofar as spiders are usually "spider enemies" and "only like to eat each other." However, the juvenile female tolerates her male companion. Shortly after the final female molt, the male copulates with her and guards her for several more days, probably to keep other competitors away from her. Then it is time for the male to take off. The expectant mother becomes increasingly aggressive toward her husband and sometimes even eats him.

Badumna longinqua prefers warm, rather humid areas all year-round with cooler retreats. The species was transported by humans to North and South America, South Africa, New Zealand, and Japan, where it now forms stable populations around human dwellings (Fig. 9.2). In New Zealand, it has probably even become the most common spider. It has also been introduced to Europe at least twice: first to Berlin in Germany, where it has not yet been able to establish itself, and then to England, where it is now found in several locations across two cities. But this seems to be far from the end of its range expansion.

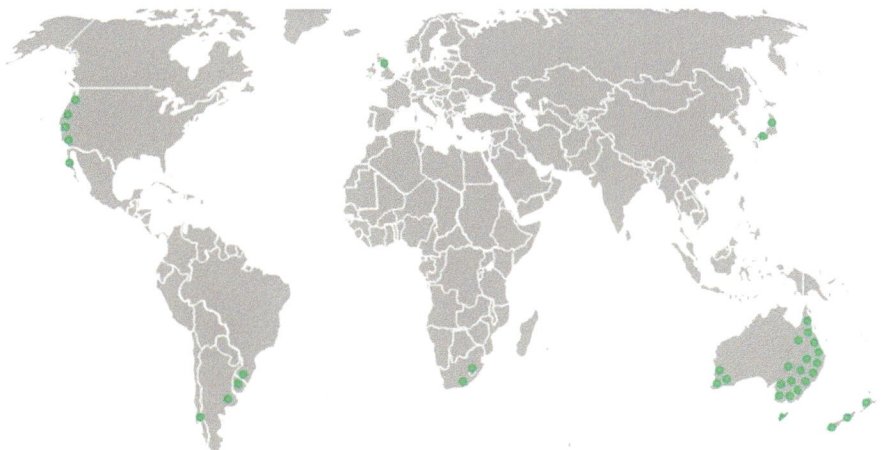

Fig. 9.2   Distribution area of the gray house spider *Badumna longinqua* (green)

Dictynidae, with around 500 species, is a globally distributed family of predominantly small spiders. Mesh-weavers belong to the **cribellate** spiders, which incorporate mesh threads (Chap. 1) into their small webs. These webs usually appear unstructured and are mostly deployed on plants, often around the spider's retreat.

10.1 Wall Spider *Brigittea civica* (Mesh-Weavers, Dictynidae)

During a stroll through a Central European city, many passersby may wonder where the numerous "mold spots" on many house walls and in underpasses come from. Upon closer inspection, you may notice that many of these spots contain an inconspicuous gray-brown spider: the wall spider *Brigittea civica* (Fig. 10.1). Over time, street dust gets caught in their small webs, which are 7–10 cm in diameter, leading to their characteristic appearance, "as if a dirty tennis ball had been thrown against the wall many times," as a German spider researcher once expressed it (Fig. 10.2).

This small spider, only 2.3–3.5 mm in size, like many synanthropic species, naturally inhabits rock walls. Its front body is reddish-brown with rows of thick, white hairs. The legs are brown, annulated whitish, and the abdomen is yellowish with a brown heart spot and paired brown spots.

This spider originally came from warmer climes and is now found in large parts of Europe, as well as in North Africa, Turkey, and Iran, and it has been introduced in South Africa (Fig. 10.3). As it is very warm-loving, its occurrence in temperate latitudes is limited to **house walls**. If these walls are also reasonably protected against wind and rain and have a rough surface with cavities and cracks, it feels particularly at home. However, it has occasionally also been found on completely smooth marble walls.

The web is a circular, usually multilayered, cribellate **capture web.** It is equipped with a catching wool made of hundreds of the finest silk fibers, which hold smaller insects extremely efficiently. The web is initially spun as a single layer. When the cribellate catching wool becomes dusty, additional layers are added, so the web can

© Association for the Promotion of Spider Research 2024
W. Nentwig et al., *House Spiders - Worldwide*,
https://doi.org/10.1007/978-3-031-70448-2_10

Fig. 10.1 Female (**left**) of the wall spider *Brigittea civica* and male (**right**) of *Dictyna calcarata*. (*Photos* left Dragiša Savić, right Paula Cushing)

Fig. 10.2 Webs of the wall spider *Brigittea civica*: **above** on a historical facade; **below** on modern exposed concrete. (*Photos* above Wolfgang Nentwig, below Pierre Loria)

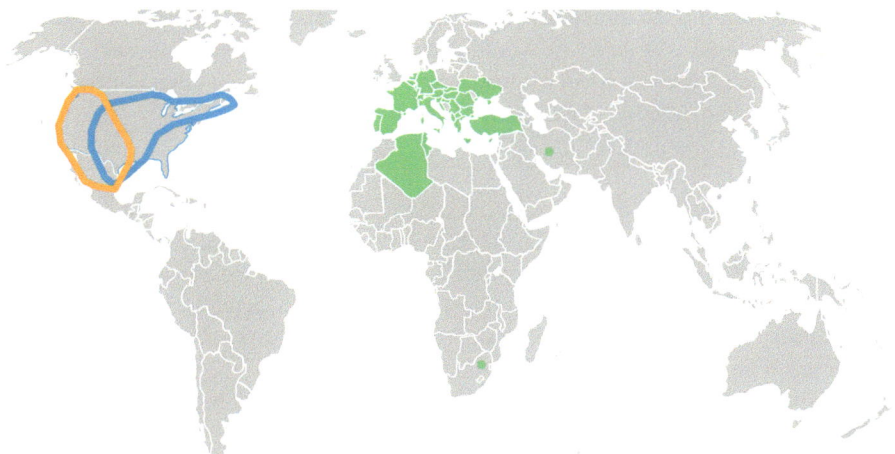

Fig. 10.3 Distribution areas of the wall spider *Brigittea civica* (green), *Dictyna calcarata* (yellow-brown), and *Dictyna bellans* (blue)

become up to 3 mm thick. However, there comes a point when the dirty net is abandoned by the spider.

Interestingly, *Brigittea civica* almost always builds two webs a few centimeters apart, which are connected by individual threads. When the first web becomes too heavily soiled by street dust, the animal retreats to the previously spun second web. Even an old female, who has already laid her eggs, migrates to the second web two to eight days before the young hatch. The mother's first web is then often taken over by one of her offspring.

Remarkable is the fact that *Brigittea civica* can also catch insects without its web. If it perceives the sound of buzzing insect wings nearby, it immediately orients itself, leaves its nest, and attacks without hesitation as soon as the insect lands on the wall. In doing so, it does not see the insect because its eight eyes are not very acute (Fig. 4.19), but it perceives the air vibrations caused by the insect wings using long, slender sensory hairs on its legs: the trichobothria.

The main prey of the wall spider are small insects, such as fruit flies or aphids. However, if the spider is very hungry, it also attacks insects up to the size of a housefly, which it almost always manages to overpower. In doing so, the spider bites somewhere on the body of the fly, often being dragged by it for several centimeters. The fly, in turn, is severely restricted in its mobility due to its entanglement in the elastic, cribellate catching threads. Through repeated venom injections from the spider's jaws, the fly is eventually weakened to the point that it can be overpowered. In one observed case, a battle between a wall spider and a housefly lasted a whopping 12 h!

Females, after mating, spin between one and three **eggsacs**, each containing between 8 and 28 eggs. The eggsacs are either deposited in cracks in the masonry or placed in a small, specially spun silken hiding place.

Brigittea civica is an unusually peaceful animal by spider standards. For example, if it has caught a large fly, it would certainly share this abundant meal with its neighbors. Such behavior is very unusual among spiders. Up to three animals have been observed sucking on a fly together.

Only in two situations do the spiders not tolerate any nonsense: females, who have laid eggs, fiercely defend their web against other females, with the web owner always winning. Males, who have found a partner, stay with her from mating until a few days before egg laying. The males guard the female and drive away other males from the common web who want to try their luck. However, this battle is more of a test of strength: the two males face each other head-on and try to push each other out of the web with their palps. Apparently, the owner of the web always wins, even if he is smaller than the intruder.

One could describe this sexual behavior of the wall spider as monogamy. Females lay their eggs a few days to weeks after copulation and die a few weeks later. The shared time of both sexes is therefore limited to only a few weeks and can also be understood as mate guarding to ensure the paternity of the male.

10.2 *Dictyna bellans* and *Dictyna calcarata* (Mesh-Weavers, Dictynidae)

Two species of the genus *Dictyna* are relatively common on houses in North America: *Dictyna bellans* and *Dictyna calcarata*. In both species, females are 2–4 mm long, and males are 2–3 mm. The front body of *Dictyna bellans* is orange to brown, the legs are yellow-orange, and the hind body varies from whitish to gray with a brown heart spot and rows of connected dots. *Dictyna calcarata* shows rows of white, thick hairs on its brown front body, the brown legs are annulated with whitish-yellow, and the light hind body has a brown heart spot and paired brown spots covered with thick white hairs (Fig. 10.1).

These species usually live outside the house in small hiding places in the corners of windows, in the cracks of house siding, or in the recesses of bricks. From these retreats, the cribellate catching threads extend to the adjacent area. As with the European *Brigittea civica*, these nets are often dusty or covered with remains of prey animals. These small spiders mainly live on the insects that are attracted by the outdoor lighting or the light from windows. Their behavior is probably similar in many ways to that of *Brigittea civica*. *Dictyna calcarata* is more common in the west of North America, while *Dictyna bellans* is more common in the east (Fig. 10.3). Both species have significantly expand their habitat into human settlement areas.

Woodlouse hunters are a spider family that today comprises a little more than 600 species in 25 genera. They have only six eyes, which are grouped at the front of the body (Fig. 4.12). The front and hind body are elongate. They are small to medium-sized species that occur in many habitats, where they are mostly nocturnal hunters on the ground, in leaf litter, under stones, and in caves of all kinds. They do not build capture webs. The family as such is distributed worldwide but has a focus in the Mediterranean area. Although many dysderids, being cave dwellers, seem suited to colonize human buildings, only one species is common in houses. It has therefore also achieved a worldwide distribution. We describe this species in more detail here.

11.1 The Large Woodlouse Hunter *Dysdera crocata* (Woodlouse Hunters, Dysderidae)

Dysdera crocata from the dysderid family is a very common house spider. The species originally comes from the Mediterranean part of Europe but has now spread across almost the whole world. Today, it is common in North and South America, all of Europe, Asia, and even in Australia and New Zealand. How is this possible?

This attractive species has a deep rust-red front body, uniformly rust-red legs, and a grayish-yellow hind body (Fig. 11.1). The jaws are quite large, strong, long, and protrude far in front of the spider. At their tip are long, thin, slightly curved fangs. The body length of females is 11–15 mm, while males reach 9–10 mm. The spider has only six eyes, which are located in the middle of the front body, are quite small, and can only be counted using a magnifying glass (Fig. 4.12). Adult spiders can live two to three years, although most other dysderid species are annual.

In North America and the English-speaking parts of Europe, this spider is commonly referred to as the "woodlouse hunter," which refers, of course, to the woodlice that are common in damp cellars and gardens. With its very long mouthparts, it

W. Nentwig et al., *House Spiders - Worldwide*,
https://doi.org/10.1007/978-3-031-70448-2_11

Fig. 11.1 Female (**left**) and male (**right**) of the large woodlouse hunter *Dysdera crocata*. (*Photos* by Gabor Kovacs)

was assumed that *Dysdera crocata* specialized in hunting woodlice. Studies have shown that these spiders, with their large jaws, can indeed grasp and pierce **woodlice** well, but in spite of their common name, woodlice are not necessarily the preferred prey of these spiders. In other words, *Dysdera crocata*, when given the choice, may choose a tastier type of prey. In the case of food shortage (which probably reflects their real situation most of the time), it is able to overpower woodlice. Most other spiders cannot overpower woodlice because of their strong, calcified exoskeleton, which acts like protective armor.

Dysdera crocata is a nocturnal hunter that hides during the day under stones or wood in a retreat spun with silk. These spiders are mainly found in damp or wet microhabitats – including mulch piles, under plant pots in the garden, in basements, utility rooms, garages, and under leaf litter – basically everywhere where humidity is high. Since the large woodlouse hunter also likes to use wooden pallets or other means of transport as a daytime hideout, this explains how it was able to conquer all continents in about 200 years (Fig. 11.2). People increasingly transported all kinds of goods and thus, among other things, also introduced *Dysdera crocata* as a stowaway.

Although the bright red color pattern and the large jaws of this spider give it an imposing appearance, the **venom** of the large woodlouse hunter is harmless to humans. If the spider is caught by hand and feels threatened, it can nip the attacker a little; otherwise, bites from this species are without consequence.

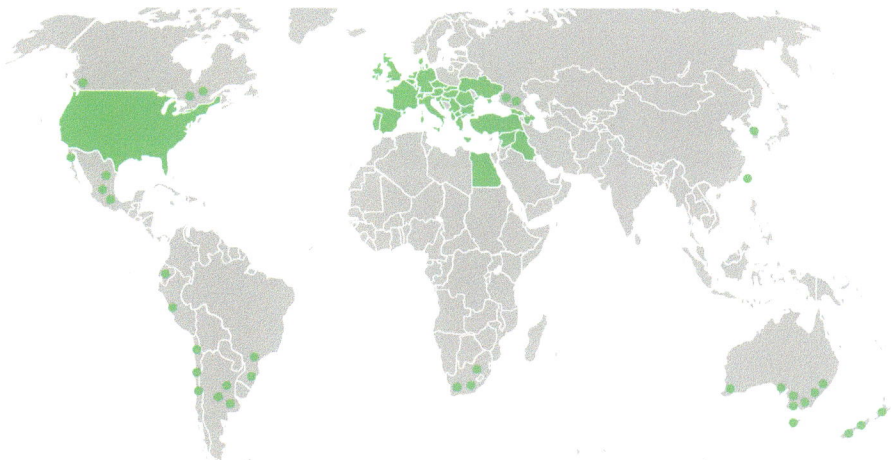

Fig. 11.2 Distribution area of the large woodlouse hunter *Dysdera crocata* (green)

Crevice weavers, Filistatidae, are a small family of nearly 200 species worldwide. They are very well adapted to dry habitats; thus, they are mainly found in deserts and semi-desert regions of the world. Since buildings in human settlements have a certain climatic similarity to the original habitat of Filistatidae, we now find them throughout the world on walls, especially in frost-free areas.

Filistatidae are small to large (20 mm body length) spiders. The front body is narrow in the area of the eight eyes, elongated, oval, and flat behind it. The eyes are arranged in a compact eye field and are not noticeably large (Fig. 4.13). The long legs, like the whole body, are covered with short hairs, giving the spiders a velvety appearance. While males are usually annual, adult females of the larger species can live for about 10 years. They continue to molt regularly as adults. This is a remarkable exception among spiders, as otherwise only tarantulas and their relatives continue to molt as adults.

A **retreat**, which leads into a rock crevice, under bark, or into a hole in the masonry, forms the central part of the filistatid **capture web**. Here, the spider hides, and from this retreat, it stretches some threads over the adjacent surface. Over time, an irregular capture sheet is created, which extends more or less concentrically around the retreat. Over time, this web sheet becomes dusty or is soiled with dirt particles. The spider occasionally repairs and enhances the web but never completely renews it (Fig. 12.1). Filistatidae are **cribellate** spiders, meaning that their capture web contains cribellate threads (Chap. 1).

Crevice weavers are **nocturnal** and hide during the day in their retreat. At night, they sit at the entrance of their retreat or on the sheet web. Despite increasing amounts of dirt, a filistatid web holds insects that accidentally come into contact with it for a considerable amount of time. Movements of the prey alert the spider, which runs toward it, preferably bites it in a leg, and injects venom. In addition to many small insects, these cribellate webs also catch larger prey like houseflies, horseflies, cockroaches, beetles, and ants.

Fig. 12.1 Retreat and capture web of *Filistata insidiatrix*. (*Photo* Jean-Philippe Taberlet)

Fig. 12.2 Female (**left**) and male (**right**) of *Filistata insidiatrix*. (*Photos* Jean-Philippe Taberlet)

12.1 *Filistata insidiatrix* (Crevice Weavers, Filistatidae)

This species is, so to speak, the Old World counterpart to the next species, *Kukulcania hibernalis*. *Filistata insidiatrix* is a brownish species in which both sexes have a lighter-colored front body. However, the eye area, along the midline of the front body, and around the edges is marked with a dark brown color. The hind body is grayish (Fig. 12.2). Like all Filistatidae, *Filistata insidiatrix* also has many short hairs, which give the animal a velvety appearance. Males are smaller, have longer, thinner legs, and are overall slimmer, while females have stronger legs. Females reach a body length of up to 14 mm, while males reach up to 9 mm.

Filistata insidiatrix is a Mediterranean species that used to live in cork oak forests and shrubland, as well as in rock crevices and under stones. Nowadays, it can also be found regularly in dry stone walls and masonry of all kinds in human settlement areas. This species has been introduced to other continents several times (Fig. 12.3).

12.2 *Kukulcania hibernalis* (Crevice Weavers, Filistatidae)

This large species (females reach 19 mm body length, males 10 mm) is referred to in the United States as the southern house spider. Females are strikingly contrasting in color, with the attachment points of the legs to the front body being almost white, while the front body and the parts of the legs close to the body are, however, dark brown to black. The unpatterned hind body and the leg tips are velvety gray. Males are medium brown, with darker coloration around the eyes that extends as a crest line along the front body. The hind body and legs are dark brown near the front body, otherwise light brown and without any pattern (Fig. 12.4). Due to this brown middle stripe, male *Kukulcania hibernalis* are apparently often confused with *Loxosceles reclusa* (Fig. 23.1), although size, markings, and general appearance are very different. In addition, the males have noticeably long pedipalps.

While females typically stay in their retreat and are thus hardly noticed by humans, adult males wander around in search of mating-ready females and can then be easily encountered. *Kukulcania hibernalis* is widely distributed from the southern United States through Central America to South America (Fig. 12.3).

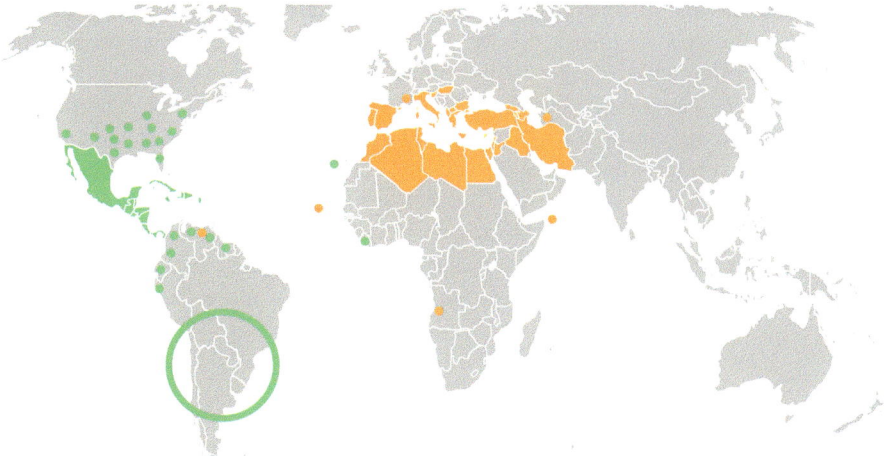

Fig. 12.3 Distribution areas of *Filistata insidiatrix* (yellow-brown) and *Kukulcania hibernalis* (green)

Fig. 12.4 Female (**left**) and male (**right**) of *Kukulcania hibernalis*. (*Photos* Gordon Ackermann)

Interestingly, most records of this species come from urban areas. Only rarely are individuals found in natural habitats, occasionally on tree trunks or rocks.

When a female has built an **eggsac**, it is stored and guarded in the area of the retreat. The young remain together until the second or third molt stage, thus showing tolerance of each other during this time. The female brings them freshly killed prey, on which they suck together. Later, the young spiders also participate in communal prey capture until they go their separate ways with increasing age.

In addition to the two species described in this chapter, there are a number of other Filistatidae species that can occur in human settlement areas. In Central and South America, these include species of the genera *Filistatinella* and *Filistatoides*. From the Mediterranean area to the inner Asian steppe regions, Iran, India, and China, species of the genus *Microfilistata* and *Pritha* occur, while in Southeast Asia, *Labahitha* is found. The lifestyle of all these spiders is quite similar, and most species are difficult to distinguish; thus, we refrain from a more detailed presentation here.

Ground Spiders (Gnaphosidae)

Ground spiders are generally poorly known, as they are nocturnal and usually live quite cryptically. They are a group of spiders distributed worldwide and, with almost 2500 species, represent the sixth largest family. They are small to medium-sized spiders, most of which are colored in brown and black tones, often with parts of the body covered in velvety hairs.

Gnaphosids have an elongate and usually somewhat flattened, robust body. The legs are all about the same length. The spinnerets of ground spiders are noticeable because they are cylindrical and usually easy to recognize from above. The eyes are small, and the rear middle eyes have an unusual oval shape for spiders, which can only be seen with good magnification (Fig. 4.24). Given the atypical whitish color of these eyes (Fig. 13.1), scientists examined them more closely. It turned out that they are specialized polarization detectors. Ground spiders can thus see polarized light in the sky, which is invisible to humans, and use this light to orient themselves. Polarized light oscillates preferentially in one plane and is present in the sky in a characteristic pattern that changes with the time of day. This light is used for orientation by many arthropods, such as honeybees. Spiders from other families (e.g., funnel-web spiders, Agelenidae) also seem to be capable of this but are less well-studied.

Ground spiders live predominantly on the ground, in leaf litter, and between stones. They do not build capture webs but actively search for prey at night. In this sense, they are true hunters. They slowly roam their habitat for potential prey animals. Most gnaphosids are unspecialized in terms of food, thus they eat anything they can catch. However, some species have specialized in certain groups of insects, mostly ants.

© Association for the Promotion of Spider Research 2024
W. Nentwig et al., *House Spiders - Worldwide*,
https://doi.org/10.1007/978-3-031-70448-2_13

Fig. 13.1 Female of the common ground spider *Drassodes lapidosus* (**left**) and her ca. 3 cm wide residential web (**right**). (*Photos* left Jean-Philippe Taberlet, right Jutta Ansorg)

13.1   Common Ground Spider *Drassodes lapidosus* (Ground Spiders, Gnaphosidae)

The Latin name of the Common Ground Spider, *Drassodes lapidosus* (*lapidosus* = stony), already indicates that this species likes to stay under stones. However, you should generally not attach too much importance to these historical names because both Latin names and, especially, common names can be misleading. As such, the life cycle of *Drassodes lapidosus* does not take place only under stones.

Females of the common ground spider can be about 9–15 mm and males 6–13 mm in size. The overall appearance of this species is quite impressive. The body and the relatively long legs appear robust, and their color is reddish-brown without any markings or pattern. Upon close inspection, it is noticeable that the entire body of this spider species has very fine, velvety hairs. The front body and legs are more reddish-brown, the eye region and mouthparts are darker, the distal parts of the legs are light gray, and the hind body appears more gray-brown (Fig. 13.1).

Drassodes lapidosus lives in warm, open, dry habitats, such as dry grasslands, dunes, and heathlands, but also in moister areas like light forests and moors. Occasionally, it can be found in houses and apartments. It is distributed across the Palearctic, i.e., it is found in Europe, North Africa, and temperate Asia (Fig. 13.2). At least in Central Europe, it is quite common. Apparently, *Drassodes lapidosus* has not yet been introduced to North America, or at least it has not been able to establish itself there. In the Americas, and in the natural distribution area of *Drassodes lapidosus*, other *Drassodes* species can also be found in houses.

The common ground spider lives mainly in the litter layer on the ground. Due to its monotonous coloring and its small eyes, you can already guess that, like most gnaphosids, it is **nocturnal**. During the day, it hides under stones or wood in a white silken retreat, which it revisits over a long period of time. At night, it hunts freely, i.e., without building a web, in search of prey. With its front legs extended like antennae, it creeps over the ground. This typical movement is described as elegant

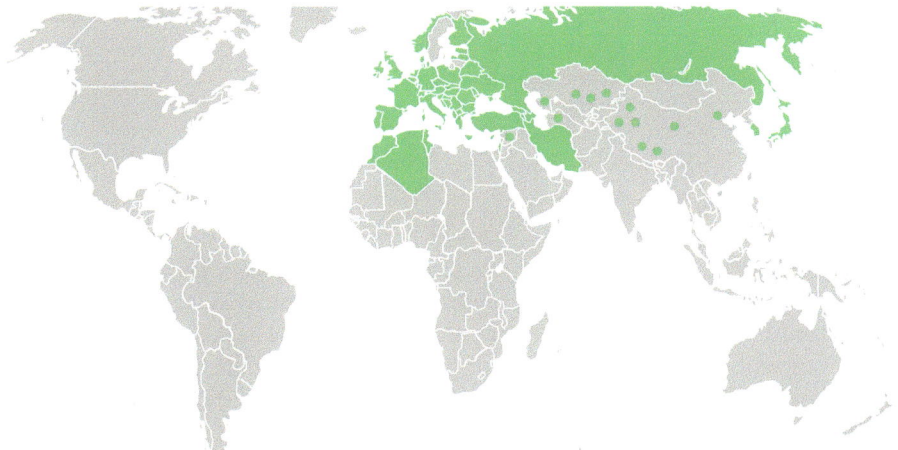

Fig. 13.2 Distribution area of the common ground spider *Drassodes lapidosus* (green)

and quick, even as "smooth as a big cat." It is also capable of moving on smooth surfaces, such as glass panes.

Smaller **prey**, such as beetles or woodlice, are attacked from the front, and the spider grabs them quickly with its venomous jaws. However, being strong and very well protected, the spider is even capable of overpowering prey that are larger than itself. It is worth mentioning that this even includes larger spiders. In this case, *Drassodes lapidosus* employs a different tactic: first, it runs past the other spider at breakneck speed, binding its prey's toe tips to the ground with a wide silk band as it passes, and then jumps on the prey's back to deliver a venomous bite. Even though this sophisticated technique leads to a larger prey spectrum and despite the fact that the common ground spider can defend itself well, it can certainly fall prey to other spiders, such as the long-bodied cellar spider *Pholcus phalangioides* (Fig. 2.1).

Mating of the common ground spider, like all ground spiders, occurs without pronounced courtship behavior. In order to avoid taking greater risks, the male prefers to find a female shortly before her last molt into a sexually mature animal. Together with the not yet fully grown female, he waits inside the cocoon for the final molt to maturity. Immediately after the molt, he mates with the female, even before her cuticle (the exoskeleton) has hardened, so that she is still defenseless. But it can also happen that two mature animals meet in the open and mating occurs.

The female builds an **eggsac** that can contain around 100 eggs. The eggsac is held by the spider between its legs and guarded inside a protective web until the young spiders hatch. Females can live up to three years. In Central Europe, they overwinter as juveniles. Some individuals overwinter in their second year as mature animals, but some still as juveniles and only undertake their final molt to maturity in the spring of their third year.

Drassodes lapidosus is not very noticeable in our homes due to its nocturnal activity, even though it is often a guest or even lives there regularly. Like almost all spiders, it is shy and flees when threatened.

13.2 Eastern Parson Spider *Herpyllus ecclesiasticus* (Ground Spiders, Gnaphosidae)

Herpyllus ecclesiasticus is commonly referred to as the eastern parson spider. This active ground spider is so named because of the attractive white pattern on the top of the hind body, which is reminiscent of a priest's tie or neckerchief. This white band often extends onto the forebody of the spider. Overall, the body of the parson spider is quite dark, except for an additional small white spot near the spinnerets on the spider's hind body (Fig. 13.3). Females grow to 6–9 mm in size, and males to 5–6 mm.

The eastern parson spider resembles the western parson spider, *Herpyllus propinquus*. These two species can only be distinguished by examining their sexual organs. Generally, the range of the eastern parson spider extends from the middle of the United States to the eastern (Atlantic) coast, while the range of the western parson spider extends from western Colorado and southern Wyoming westward to the Pacific coast and southward through Mexico. Further records of *Herpyllus ecclesiasticus* come from southern Canada (Fig. 13.4).

Adult animals can be found throughout the year. Like most species from the family Gnaphosidae, the parson spider runs quickly on the ground and actively hunts for prey. It is mainly active during the evening and night. During the day, it hides under

Fig. 13.3 Female of the eastern parson spider *Herpyllus ecclesiasticus*. (*Photo* Brandon Woo)

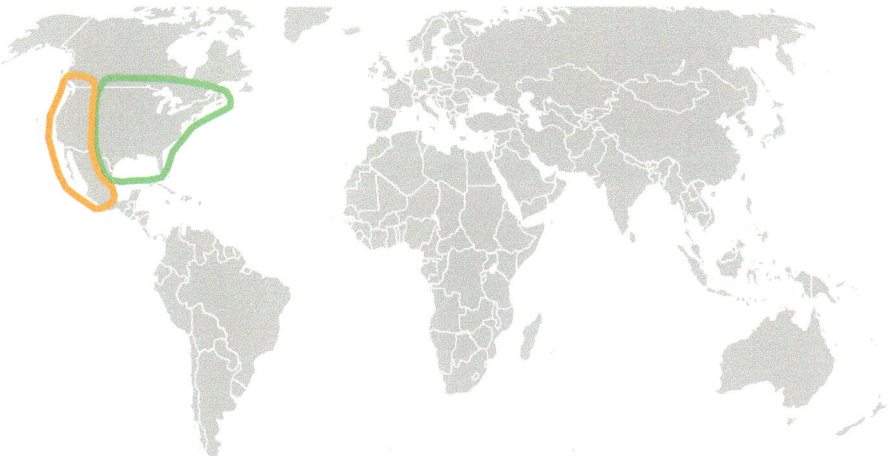

Fig. 13.4 Distribution areas of the eastern parson spider *Herpyllus ecclesiasticus* (green) and the western parson spider *Herpyllus propinquus* (yellow-brown)

stones, tree trunks, and other debris on the ground. It has been found in the mountains at altitudes of up to 2450 m. The female builds a flat **eggsac**, places it under bark or other structures, and guards it. It has been reported that *Herpyllus* species also eat dead insects, thus they are partially scavengers. Another very special behavior of a *Herpyllus* species is that it defends itself by dropping a wide strip of silk from its spinnerets to protect itself from an attack from behind. It is not known whether these strange behaviors for spiders apply to all currently known 31 species of the genus *Herpyllus*.

The eastern parson spider sometimes invades houses and can also be found near houses under objects lying on the ground. It is unlikely that these spiders will come into closer contact with humans or even bite them, as these rather fast runners always scurry away over the ground.

13.3   The Spotted Mouse Spider *Scotophaeus scutulatus* (Ground Spiders, Gnaphosidae)

Resembling a small dark spot on the wall, the spotted mouse spider (*Scotophaeus scutulatus*) stealthily moves forward in search of prey. It repeatedly pauses, then quickly pounces on mosquitoes, moths, or other small insects. At 6–15 mm, it is one of our medium-sized cohabitants. The forebody, mouthparts, and legs are reddish-brown. The hind body appears rather grayish-brown and is densely covered with silvery shiny hairs, giving the spider a fluffy, almost woolly appearance such that it resembles a small mouse (Fig. 13.5). The spinnerets are cylindrical and strikingly long. The eyes appear rather small in relation to the forebody.

Fig. 13.5 Female (**left**) and male (**right**) of the spotted mouse spider *Scotophaeus scutulatus*. (*Photos* Pierre Loria)

Males possess a small hardened plate (the shield or scutum) on the hind body, which is only visible upon close inspection and would be difficult for most observers to see. The spotted mouse spider is usually only active at night and spends the day hidden between cracks and crevices, behind pictures, or in the folds of curtains. It can go for months without water and is thus well equipped for life in human dwellings. Females build thick-walled eggsacs, which they hide well, so that the eggs are excellently protected.

The scientific name *Scotophaeus scutulatus* indicates both the lifestyle of the spotted mouse spider (*scotos* is Greek and means darkness) and its appearance (*scutulatus* is Latin and means "with a small shield"). Additionally, two other very similar *Scotophaeus* species occur in buildings. The small mouse spider, *Scotophaeus blackwalli*, lives under the bark of trees and tree stumps, as well as in buildings. The four-dotted mouse spider, *Scotophaeus quadripunctatus*, lives in caves and also within buildings.

Scotophaeus scutulatus is found from Europe to Central Asia (Fig. 13.6) and is less common in the wild than in buildings. *Scotophaeus quadripunctatus* and *Scotophaeus blackwalli* also have a European distribution, with *Scotophaeus blackwalli* additionally being introduced into North and South America (Fig. 13.6).

13.4 House Ground Spider *Urozelotes rusticus* (Ground Spiders, Gnaphosidae)

The small house ground spider is rather inconspicuous. It usually only shows itself at night or is found, to the owner's surprise, in the sink or bathtub in the morning. It shares this behavior with the giant house spider *Eratigena atrica* (Chap. 5.2), as both spiders have no adhesive hairs on their legs to free themselves from awkward situations involving smooth walls. The cryptic lifestyle is probably the main reason why little is known about this spider.

Females reach a body length of up to 10 mm, while males are around 7 mm. The front body is medium brown to reddish brown, often appearing orange; the hind body is whitish gray to medium gray and covered with black hairs; and the legs are

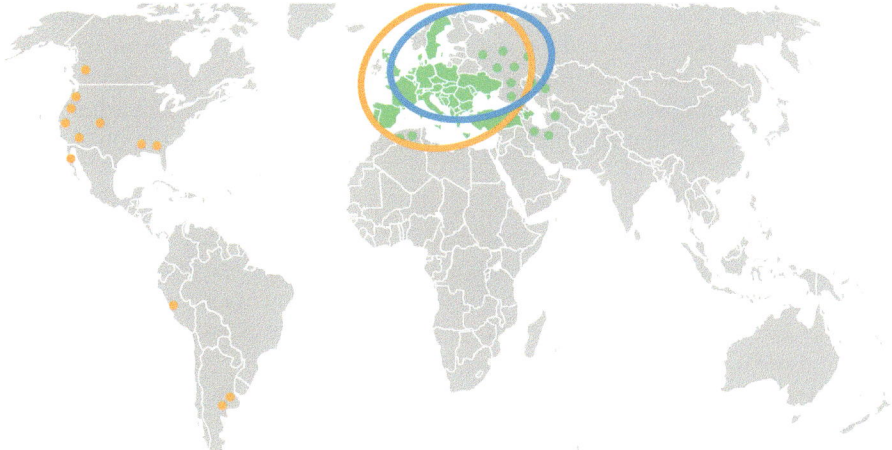

Fig. 13.6 Distribution areas of the spotted mouse spider *Scotophaeus scutulatus* (green), the small mouse spider *Scotophaeus blackwalli* (yellow-brown), and the four-dotted mouse spider *Scotophaeus quadripunctatus* (blue)

Fig. 13.7 Female of the house ground spider *Urozelotes rusticus*. (*Photo* Dragiša Savić)

orange-brown and also very hairy (Fig. 13.7). The front pairs of legs bear spines. The spinnerets are long and cylindrical, and the eyes are small and oval, as is usual with gnaphosids. It is difficult to distinguish this species from its close relatives; to do so, their sexual organs would need to be examined.

Urozelotes rusticus runs jerkily and very quickly over the floor, then suddenly sits still with its legs folded. On a dark background, it becomes almost invisible. Nothing is known about the prey of the house ground spider. However, we can assume that it captures a wide range of small insects that can occur in our houses or apartments. It thus eliminates many pests and is an ideal roommate in our homes. Outdoors, it likes to hide under stones and everything that lies on the ground. Females produce flat, almost disk-shaped whitish or pink eggsacs.

Urozelotes rusticus is a spider that has long puzzled researchers. It is found in houses today throughout the world, although it is not at all clear where it originally came from (Fig. 13.8). The idea that this spider, like many others, could have been passively spread from its area of origin in all directions across the world with human cargo did not occur to early spider researchers. Therefore, this species was **repeatedly described** as a separate new species, albeit from very different locations. This happened at least 26 times over 100 years. It was only later noticed that all these descriptions actually referred to the same species, *Urozelotes rusticus*.

Today we assume that the house ground spider most likely originates from the Mediterranean region and has so far been introduced to all continents except Australia. Its lifestyle of hiding under anything that lies on the ground has naturally promoted its spread. Thus, it probably went from a warehouse to a ship, which sailed to a distant port, then into another warehouse, and was subsequently transported within the port city or to neighboring cities. In a little more than 100 years, it had traveled around the whole world. Today we also know that the house ground spider has been found in gardens, pastures, orchards, forests, and caves. It therefore seems to have few demands on its environment, which certainly helped it conquer the world.

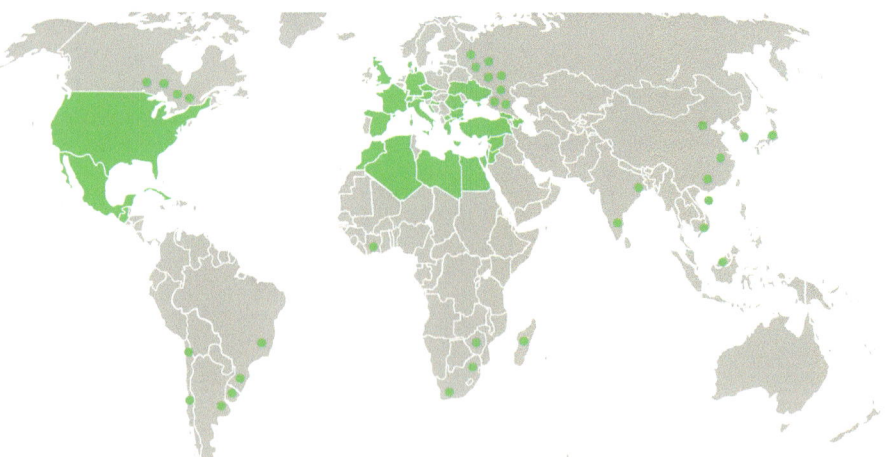

Fig. 13.8 Distribution area of the house ground spider *Urozelotes rusticus* (green)

Sheet-Weavers (Linyphiidae)

Linyphiidae comprise nearly 5000 species worldwide, making them the second most common spider family after the jumping spiders. They are web-building spiders with a body length of 1–8 mm (although most species are only 1–3 mm), which are recognizable by their horizontal **sheet webs**. Ground-dwelling linyphiids build their tiny sheet webs in crevices in the ground, which can be seen early in the morning with dew covering the silk. Aerial webs built by spiders in this family often consist of a horizontal or dome-shaped web sheet, frequently called the canopy due to the way it curves downward at the peripheries. In some species, the web sheet is shaped more like a cup or container; these webs are often referred to as "bowl and doily" webs because of their shape. With a system of suspension and alarm threads, the web sheet is held under tension both above and below and is anchored to the vegetation. Contact with these threads causes approaching animals to create **vibrations**, which are transmitted to the web sheet. The spider thus receives information about whether it is a potential prey or a predator, allowing it to react accordingly: attack or escape. The eight small eyes, on the other hand, do not facilitate visual orientation (Fig. 4.22).

Sheet-webs are intended as permanent installations. The spider therefore regularly cleans them of dirt and repairs damaged parts. The new web often does not contain any adhesive droplets for catching prey. However, many linyphiid species use them for repairing the web. Normally, sheet-weavers sit underneath the canopy and lurk there waiting for prey, with their bellies upwards and their backs downwards. If an insect gets caught in the upper tangle of threads, the spider shakes the web vigorously until the prey falls onto the web sheet. Often, the spider runs up the edge of the canopy and overpowers the prey. Sheet-weavers do not have a retreat or any similar place to run to. When threatened, they flee into the adjacent vegetation or drop to the ground.

Linyphiidae occur throughout the world in almost all habitats but prefer the vegetation of temperate latitudes. Especially in the morning dew, their three-dimensional webs are often seen in large numbers due to the dew collecting on them. Many

© Association for the Promotion of Spider Research 2024
W. Nentwig et al., *House Spiders - Worldwide*,
https://doi.org/10.1007/978-3-031-70448-2_14

sheet-weavers have modified and reduced their webs. Small species, in particular, often build catching sheets only a few centimeters across with individual suspension threads.

Species of Linyphiidae are a very difficult group to identify. In many habitats, there are many different species, most of which are only small in size, and the only sure features to distinguish them are their sexual organs. A good binocular microscope is therefore indispensable if you want to study Linyphiidae more closely.

Linyphiidae are not actually house spiders per se, but they often occur outside buildings in high species numbers and with high individual density. Thus, it should not be surprising that a linyphiid is seen in the house every now and then. However, it has almost always lost its way into the building and cannot survive indoors in the long term. By far, the most common sheet-weaver species found in houses is *Lepthyphantes leprosus*, which we therefore describe in more detail here. In addition, other *Lepthyphantes* species occasionally occur, as well as some other Linyphiidae species. Again, they accidentally enter houses, cannot survive there, and are not considered further here.

Fig. 14.1 Female (**left**) and male (**right**) of *Lepthyphantes leprosus*. (*Photos* left Guido Gabriel, right Michael Hohner)

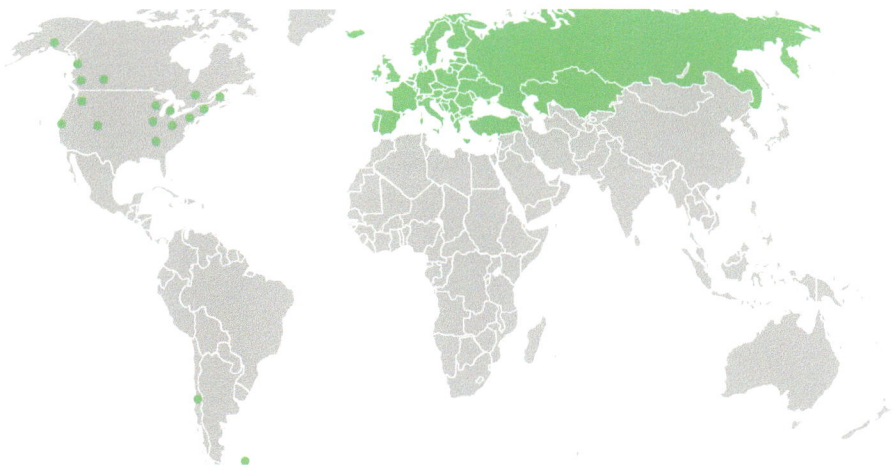

Fig. 14.2 Distribution area of *Lepthyphantes leprosus* (green)

14.1   *Lepthyphantes leprosus* (Sheet-Weavers, Linyphiidae)

This is a 2.5–4 mm large spider species. The front body is light brown to brown; sometimes the edge is dark or a dark radial pattern of stripes can be seen. The legs are light brown to brown, often with dark annulations. The rear body is yellowish-gray on top with dark markings (Fig. 14.1).

Originally, *Lepthyphantes leprosus* probably lived between stones, under bark, and in caves. Today, it is often found in buildings, especially where it is damp. These places include basements, laundry rooms, stables, tunnels, mines, and sewer shafts. Adult animals are present all year round. The spider builds its inconspicuous web, in which you can distinguish a fine web sheet and some suspension threads, primarily in corners and next to edges or protrusions.

The original distribution area of *Lepthyphantes leprosus* was Eurasia. This species was then probably transported with sailing ships to North America, where it spread along both coasts and in the area of the Great Lakes. The species has also been found in Chile and the Falkland Islands but is likely to be even more widespread, especially in cool, damp regions (Fig. 14.2).

Pirate Spiders (Mimetidae)

15

Mimetids are an exciting spider family of only 160 species, but they present a highly unusual prey specialization. These are rather small species that occur throughout the world but have a clear distribution focus in the tropics. A hint to their lifestyle is the conspicuous spikiness of the forelegs, which can be used to capture particularly dangerous prey. As the English name "pirate spider" already suggests, mimetids have specialized in attacking and plundering things: namely, other spiders' webs and the spiders that live within them.

15.1 Four-Humped Pirate Spider *Ero aphana* (Pirate Spiders, Mimetidae)

This spider is rarely seen and yet is a common cohabitant on house walls in European cities: the four-humped pirate spider, *Ero aphana* (Fig. 15.1). Its inconspicuousness is because the small spider, when disturbed, pulls its legs close to its body and hangs motionless on a thread like a gray, inconspicuous lump. The forebody is yellowish-white with a conspicuous black marking, the legs and the hind body have a yellow-brown to greenish-gray speckled camouflage pattern, and the legs are often annulated. Also worth mentioning are the name-giving humps on the hind body. The small eyes are quite close together (Fig. 4.6). Females are 3–4.5 mm long, and males are just under 3 mm.

More commonly seen than the pirate spider itself is the beautiful **eggsac** of this species (Fig. 15.1c). It is pear-shaped and suspended on a string of spider threads. The eggs lie in a light, fine tissue. On the outside, the eggsac is covered by stronger, reddish-brown, shiny, and wavy threads, which are thought to prevent ants from eating the eggs.

Ero aphana occurs from the Macaronesian Islands across Europe, North Africa, and Turkey into Central Asia. It was introduced by humans into Australia, China, and Japan, as well as on several islands. In Europe, the four-humped pirate spider has expanded its originally rather southern range toward the north in recent decades

© Association for the Promotion of Spider Research 2024
W. Nentwig et al., *House Spiders - Worldwide*,
https://doi.org/10.1007/978-3-031-70448-2_15

Fig. 15.1 The four-humped pirate spider *Ero aphana*: (**a**) female, (**b**) male, (**c**) eggsac. (*Photos*: (**a**) Pierre Oger, (**b**) Pierre Loria, (**c**) Dragiša Savić)

(Fig. 15.2). It likes to inhabit warm, dry pine forests, where the species can be found on branches and twigs of the Scots pine (*Pinus sylvestris*). From this natural habitat, *Ero aphana* has migrated to the cities – but why? This has to do with its **prey animals**: mimetids almost exclusively eat other spiders, and the favorite food of the four-humped pirate spider is cobweb spiders (Theridiidae, see Chap. 25). Occasionally, sheet-weavers (Linyphiidae) and orb-weavers (Araneidae) are preyed upon too (Chaps. 14 and 7). The numerous cobweb spiders on house walls offer the

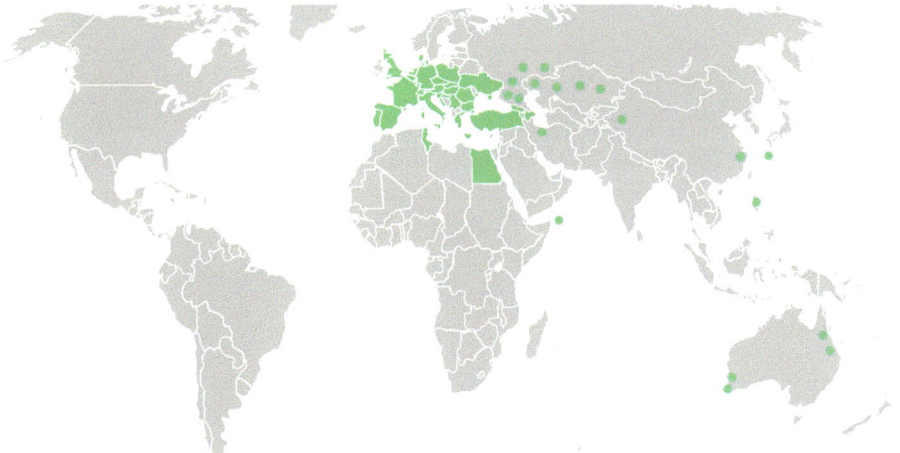

Fig. 15.2 Distribution area of the four-humped pirate spider *Ero aphana* (green)

pirate spider rich pickings. And in any case, the climate in our cities has become more Mediterranean in recent decades.

Pirate spiders usually move relatively slowly and leisurely. Once *Ero aphana* has discovered a cob-weaver's web, it stealthily approaches like a cat, without being noticed by the web owner. Then the pirate carefully removes threads that are in its way, creating a thread-free zone in front of it. Later, typically at dusk, it begins to pluck at the web, causing the web owner to approach "curiously." The purpose of this lure of the prey spider is still not entirely clear. Presumably, through its plucking pattern, the four-humped pirate spider imitates a prey insect that has become entangled in the web. However, some researchers believe that the spider imitates the mating pattern of a male spider that is ready to mate and, in this way, lures the spider it is going to prey upon.

In any case, with its front legs extended, the pirate spider waits until the cobweb spider has come close enough. Then it lunges forward quickly, pulling the victim toward it with its spiny front legs, bites the cobweb spider in a leg, and injects its highly potent venom. Astonishingly, the prey spider then collapses almost instantly. If it drops as a flight reaction, it is caught by the long spines of the pirate spider's front legs. These spines thus form a kind of "catching basket" to embrace the victim.

This hunting strategy is not without risk: sometimes the pirate spider itself becomes the prey of a cobweb spider. Especially young pirate spiders have often been found sucked dry in the webs of cobweb spiders.

Spider eaters are surprisingly peaceful spiders with regard to each other. They can easily be kept together in a glass, which is actually atypical for spiders. Mating is also peaceful: the female, when ready to mate, sits on some threads, which the male causes to vibrate "seductively" with his pedipalps. He also spins his own copulation thread toward the female, which he sets into lively movements using his second and fourth pairs of legs. Finally, the female moves a little toward the male, and mating occurs. Sexual cannibalism, i.e., the eating of the male by the female during or after mating, has never been observed in the four-humped pirate spider.

Disk Web Spiders (Oecobiidae)

16

Disk web spiders are a small family of 6 genera with only 125 species, which have a distribution focus in the tropics but are also regularly found in the subtropics. Three-quarters of these species belong to the genus *Oecobius*. They prefer to live under stones but also on vertical structures such as rocks and tree bark, and they have many adaptations or abilities that enable them to live on the walls of buildings in human settlement areas.

Species of the genus *Oecobius* are, with body lengths of less than 3 mm, small spiders that are often hardly noticed. Nevertheless, they are easy to recognize because their forebody is round and the eyes are close together on an eye mound (Fig. 4.14). The legs are short. The hind body is teardrop-shaped and slightly flattened. They are **cribellate** spiders, but this is difficult to recognize in such small animals (Chap. 1).

Oecobius species spin a **tent-like retreat**, with a size of up to 10 mm, which is stretched and secured in all directions like a tassel. Therefore, it often also appears star-shaped and has exits in different directions (Fig. 16.1). On the substrate, long **signal threads** a few centimeters long are stretched radially around the retreat. The spider stays on the bottom of the tent, which is also spun from silk threads, and thus feels any vibrations when prey approaches.

When *Oecobius* perceives through **vibrations** that a suitable prey item has touched one of the signal threads, it rushes out of its retreat, hurries to the prey, and encircles it at high speed. During this action, it raises its hind body and produces a wide band of threads with which the prey is bound. On the one hand, the legs, antennae, and mouthparts of the prey are secured in this way. On the other hand, the whole prey animal is also fixed to the ground. Thus rendered defenseless, the spider looks for a protruding leg or an antenna and applies its venomous bite there. This unusual prey-catching behavior, in which the spider whirls around its prey, means that sometimes it is also called the whirligig spider.

After being captured, the prey is also sucked dry via this leg or the antenna. For this reason, the wrapped-up and hardly chewed insects that make up the spider's diet can be easily analyzed. Prey often consists of small flies. Ants are also captured,

© Association for the Promotion of Spider Research 2024
W. Nentwig et al., *House Spiders - Worldwide*,
https://doi.org/10.1007/978-3-031-70448-2_16

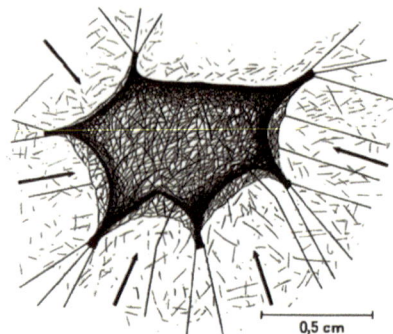

Fig. 16.1 The tent-like retreat of the disk web spider *Oecobius*. **Left** as a drawing, with entrances marked with arrows; **right**, an actual web on a building next to a busy road. (*Figure* left Lothar Glatz (1967), *photo* right Wolfgang Nentwig)

which can be dangerous opponents and defend themselves against spiders. Through its careful approach tactics and wrapping behavior with spider silk, *Oecobius* species have more or less developed into ant specialists.

16.1   Citrus Disk Web Spider *Oecobius navus* (Disk Web Spiders, Oecobiidae)

The basic color of the strikingly circular forebody and legs of the citrus disk web spider is light beige. On the edge of the forebody, there is usually a black edge marking and some spots at the top. The eye mound is typically marked dark brown to gray-black, and sometimes there is a dark midline extending over the entire forebody. The legs show a dark, annulated pattern, which can be very striking but is sometimes only indistinctly developed. The hind body shows a pattern of white guanine spots and black dots on a beige background (Fig. 16.2). **Guanine** is a degradation product that is usually excreted with the spider's feces, but here it is also stored in the form of crystals in converted intestinal cells. Through the thin cuticle, the crystals become visible as white spots, and this color and pattern camouflage the spiders extremely well on stony ground or against tree bark. These spiders are easily overlooked, which of course is also due to their small size. Females are only 2.5–2.9 mm long, and males are 2.0–2.6 mm.

Oecobius navus was originally a Mediterranean species but has been showing expansion tendencies within Europe for decades (possibly also promoted by human-induced **climate change**). Therefore, it is now also found in northern and eastern European countries, where, in climates that are too cool, it can only be found inside buildings. In addition, this spider species was apparently often spread with human goods (especially building materials). Therefore, with human help, the citrus disk web spider has conquered all continents of the Earth (Fig. 16.3).

Fig. 16.2 Female of the
citrus disk web spider
Oecobius navus. (*Photo*:
Guido Gabriel)

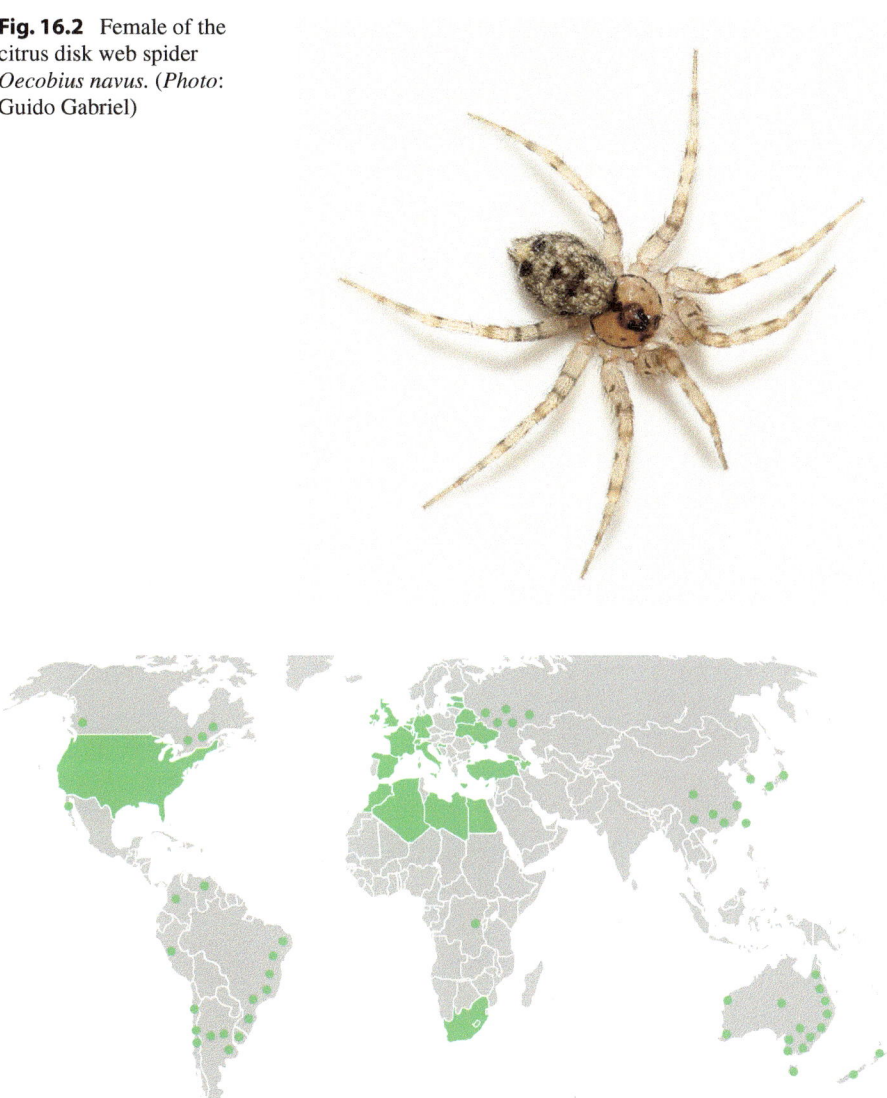

Fig. 16.3 Distribution area of the citrus disk web spider *Oecobius navus* (green)

In addition to *Oecobius navus*, other closely related *Oecobius* species are also found on human buildings. This is due to the fact that the biology of most *Oecobius* species differs little from one another. On the other hand, *Oecobius* species are difficult to distinguish due to their similarity and the variability of their color patterns. At least 12 other *Oecobius* species have occasionally been found on buildings and inside houses, and unfortunately, one or more misidentifications cannot be ruled out.

Goblin spiders are a family of tiny spiders that grow to 1–2 (rarely 3) mm in size. Currently, almost 2000 species in 115 genera are known. These short-legged spiders are predominantly nocturnal and do not build capture webs but actively hunt small insects. Oonopids are found almost worldwide but have a clear focus in the tropics and subtropics of the world. They inhabit leaf litter, the top layer of soil, and also vegetation.

Goblin spiders often have a yellowish-pale appearance. Many have spiny legs, which help to grasp their prey. Some oonopids have massively reinforced their exoskeleton, such that we can speak of a **suit of armor** or armored spiders. This armor can be spectacularly adorned with thorns in some tropical species. In females of some other species, the hind body is wedged between the two half-shells of armor, such that the spider resembles a tiny hot dog. However, due to the small size of these spiders, this is only visible with a magnifying glass. In some languages, oonopids are known as "six-eyed spiders" since members of this family only have six eyes (Fig. 4.11). Of the spider's original eight eyes, two have become reduced in the course of their evolution. Some species have lost even more eyes. For example, there are also oonopids with only two or even no eyes.

It is also fascinating that some oonopids reproduce parthenogenetically. In other words, they are unisexual, without males. Offspring arise from unfertilized eggs. Quite atypical for spiders, the egg sacs contain only a very few eggs, in some species only one or two, which are very well guarded.

17.1 House Goblin Spider *Oonops domesticus* and Other Oonopids (Goblin Spiders, Oonopidae)

Slowly and carefully feeling its way, the house goblin spider (*Oonops domesticus*) moves forward, like a goblin, in the dark at night. It runs along walls, over wallpaper, or between the pages of books. It keeps stopping, backing away, or suddenly running off, making jerky movements and changes of direction. The small, pale

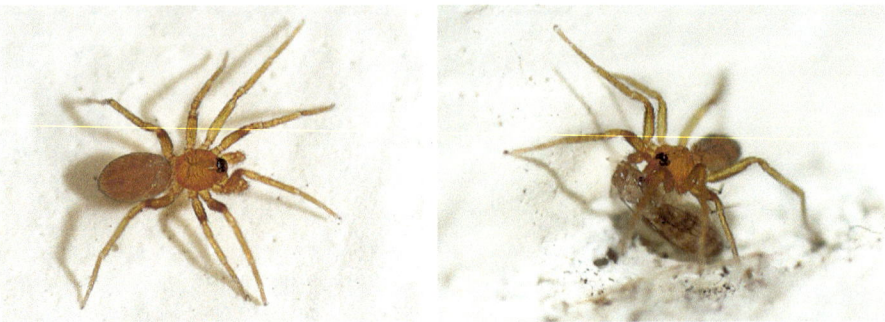

Fig. 17.1 The house goblin spider *Oonops domesticus* in ambush (**left**) and with prey (**right**). (*Photos* Eckhart Derschmidt)

body is hardly noticeable against the background. This little fellow measures only about 1.5–1.8 mm in length, and thus is about the size of a pinhead.

The first description of the house goblin spider was made in 1916 by Raymond de Dalmas. He found two female specimens in his apartment in Paris among old, dusty papers. The fore and hind bodies of *Oonops domesticus* are whitish pale, yellowish to pink, and without patterning. The legs are pale yellow, and only the front legs carry five pairs of spines in both males and females. The six eyes are closely placed (Fig. 4.11), and the eye region is almost as wide as the forebody (Fig. 17.1).

Their small size and nocturnal, cryptic lifestyle are the main reasons why these spiders are rarely seen. Their biology is largely unknown. As **prey**, they probably feed on psocids (such as book lice, which are common in many houses) and other tiny insects. Because we rarely notice these small insects, we can hardly imagine that the house goblin spider can survive off them. Reproduction occurs with several **eggsacs**, into each of which only two eggs are laid.

The various species of goblin spiders living in and on buildings are rather inconspicuous and difficult to distinguish, as they are all very small (only 1–2 mm) and pale. These include, among others, the common goblin spider (*Oonops pulcher*) in addition to the house goblin spider described above. Both species have a distribution focus in Europe (Fig. 17.2).

There are other goblin spiders that are also commonly found in houses. The swift goblin spider (*Ischnothyreus velox*) lives in the temperate zone in the litter of greenhouses. The ivy parchment spider (*Tapinesthis inermis*) is mainly found in Europe, in ivy growing along house walls. The virgin honey spider (*Triaeris stenaspis*) reproduces **parthenogenetically** (i.e., asexually, hence its strange common name) and is found in the tropics throughout the world, as well as in greenhouses outside the tropics (Fig. 17.3).

The distribution maps of these species reveal very scattered distribution areas for each species, which mainly reflect gaps in our knowledge of their actual occurrence. These very small and cryptic animals are often overlooked, even by experts. In addition, the many isolated occurrences indicate the ease with which oonopids can be spread throughout the world with the transport of goods and often survive in climatically unfavorable regions in the warm greenhouses of botanical gardens.

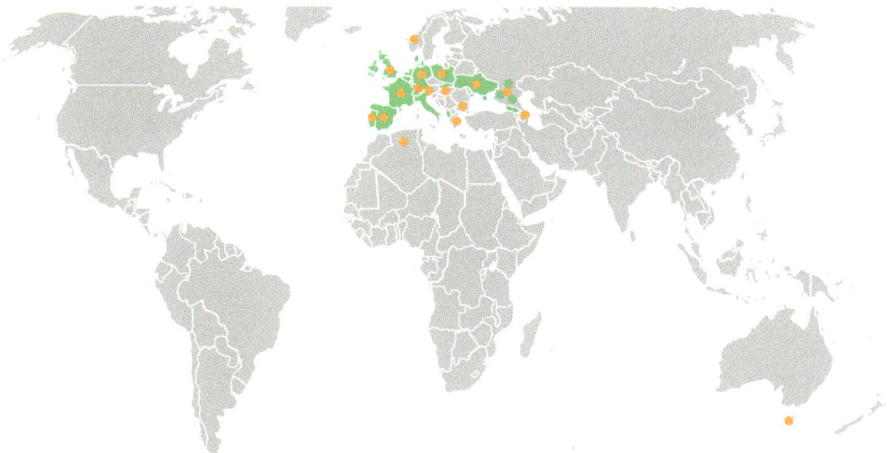

Fig. 17.2 Distribution areas of the house goblin spider *Oonops domesticus* (green) and the common goblin spider *Oonops pulcher* (yellow-brown)

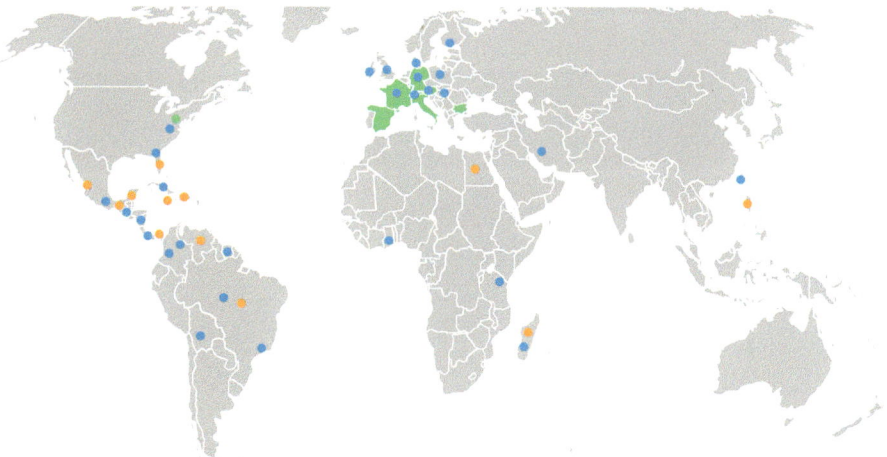

Fig. 17.3 Distribution areas of the swift goblin spider *Ischnothyreus velox* (yellow-brown), the ivy parchment spider *Tapinesthis inermis* (green), and the virgin honey spider *Triaeris stenaspis* (blue)

Cellar spiders are a family of web-building spiders found throughout the world. Currently, around 2000 species in almost 100 genera are known. Cellar spiders are easily recognized by their unusually long legs, which is why they are often confused with harvestmen. Harvestmen, however, are not spiders but represent their own order within the arachnid class, which also includes scorpions, mites, and other groups. Most cellar spiders are found in forested areas of the tropics and subtropics, where they live under stones, under roots and bark, or in tree cavities; some are also found in rock walls and stony caves.

Cellar spiders living in caves have ideal prerequisites to also conquer human buildings. They like the microclimate. In addition, they can easily be transported with human merchandise, attached as stowaways, so to speak. It is therefore not surprising that cellar spiders are now found throughout the world as regular and also frequent house spiders.

Cellar spiders include small to medium-sized species with a body length of 2–10 mm. The legs are extremely long and delicate and can be three to five times the body length, and in exceptional cases, even more. Most species are inconspicuously pale-colored (yellow, light brown, light gray) and therefore look like typical cave spiders. They build irregular and only poorly recognizable **three-dimensional webs**, in which individual threads often only become visible through dirt adhering to them. Sometimes you initially see a cellar spider hanging in the room and only later realize that there is a web around the animal.

Different prey animals get caught in cellar spiders' webs. The spider overpowers them as soon as it is alerted to potential prey via the prey-triggered **vibrations** of the capture web. Through a clever wrapping technique, spider silk is packed onto the prey with their long legs, rendering the prey completely defenseless. Cellar spiders have a broad diet that, in addition to many small insects, also includes prey items that are dangerous (ants, bees, wasps, and other spiders), hard to overpower (woodlice or beetles), or unusually large.

When threatened by an enemy, or when the whole web is shaken, some species of cellar spiders show a remarkable trembling behavior; in German, for example,

W. Nentwig et al., *House Spiders - Worldwide*,
https://doi.org/10.1007/978-3-031-70448-2_18

they are commonly referred to as "trembling spiders." During this behavior, the animals and surrounding parts of the web are set into vibrations (**trembling**), such that the spider becomes blurred in front of the enemy's eyes and can no longer be targeted. It should be noted that large cellar spiders cannot penetrate human skin with their rather small mouthparts; thus, they are completely harmless to humans.

Here, we present the seven most common cellar spiders found in houses. To decide which species is seen, it may be sufficient to flip through the following pages and look at the illustrations to find the greatest possible similarity. For species identification, it is definitely recommended to use the key provided here. For this, a good magnifying glass is needed, or even better, a binocular microscope, with the help of which you can get to know the key features. See also the explanations in Chap. 4.

1a	Six eyes in two groups of three (Fig. 4.2). Body length up to 3 mm, hind body spherical	*Spermophora senoculata*
1b	Eight eyes with very small front middle eyes (Fig. 4.2)	**2**
2a	Forebody evenly arched, without depression in the middle at the top, body length more than 4 mm, with cylindrical hind body	*Pholcus phalangioides*
2b	Forebody with depression in the middle at the top	**3**
3a	Body length 3 mm, hind body spherical	*Psilochorus simoni*
3b	Body length 5–10 mm	**4**
4a	Hind body taller than long	*Artema atlanta*
4b	Hind body longer than tall	**5**
5a	Hind body pointed upwards at the back	*Crossopriza lyoni*
5b	Hind body rounded	**6**
6a	Legs without small black dots/lines	*Smeringopus pallidus*
6b	Legs with many small, black dots/lines	*Holocnemus pluchei*

18.1   *Artema atlanta* (Cellar Spiders, Pholcidae)

This cellar spider is among the largest members of its family in the world. Females reach body lengths of 8–11 mm, males 9–10 mm, of which about 4–5 mm are for the forebody. The first pair of legs reaches about six times the body length. The forebody is white-yellowish and bears a narrow brown band at the top, and sometimes small brown spots can be seen on the sides. The legs are beige to medium brown and have brown or black annulations at three points. The hind body is whitish, with dark, irregular light gray spots, left and right of a light central stripe (Fig. 18.1).

Originally a cave spider that lives in rocky landscapes, *Artema atlanta* is the only species among a dozen other species of the genus *Artema* that has made the leap into human buildings. Here, it can be found in basements and other rooms as well as in outbuildings. As an unintentional stowaway of transported goods, this cellar spider was probably quickly spread throughout the world. It has been found, for example, on the outer walls of containers.

Fig. 18.1 Female *Artema atlanta*. (*Photo* Gordon Ackermann)

Fig. 18.2 Distribution area of *Artema atlanta* (green)

This species probably originates from an area that extends from North Africa through the Middle East to Inner Asia, although more precise details are not yet known. It then found its way to America, southern Africa, other parts of Asia, and Australia, becoming very common in many areas. In Europe, *Artema atlanta* never gained a foothold (Fig. 18.2), except in the eastern Mediterranean. Despite individual introductions to England and Belgium, the species did not establish itself there. Interestingly, this species has become rarer in some regions, such as the east coast of Brazil and on some Caribbean islands, where it has nearly disappeared. The reasons for this are not known.

18.2   Tailed Cellar Spider *Crossopriza lyoni* (Cellar Spiders, Pholcidae)

This cellar spider has extremely long legs. The front legs can be up to 6 cm long, with a body length of 4–9 mm (females) and 3–7 mm (males). In addition, the white and black annulations at two of the joints on each leg are noticeable. The forebody is almost circular, light brown with a brown stripe in the middle, and at the top in the middle, there is a dent. Eight eyes sit at the front edge (Fig. 4.2). The hind body is particularly striking, as it is abruptly blunted at the upper rear end, is longer than it is high, and appears trapezoidal in side view. It can vary greatly in color, usually light gray with dark lines or spots, and sometimes with white stripes (Fig. 18.3).

Crossopriza lyoni builds extensive **sheet webs** in all kinds of buildings. These webs are cleaned of the coarsest dirt as necessary. However, if they become too dirty, the spider builds a new web nearby. Over time, especially when they occur in larger groups, they can fill extensive parts of a room or even a warehouse with their webs, such that this negatively affects the human co-inhabitants of the buildings. These pholcids are therefore often perceived as annoying and unhygienic. Since the regular removal of the webs (for example, with a broom or vacuum cleaner) costs time and money, they are also often classified as pests.

A counter-argument to this is that *Crossopriza lyoni* (like most pholcids) catches a large number of **mosquitoes** and other disease carriers. Up to 20 mosquitoes per day have been counted in just one web. Even young spiders at their second life

Fig. 18.3 Female of the tailed cellar spider *Crossopriza lyoni*, **left** from the side, **right** from above. (*Photos* left John W Friel, right Gilles Arbour/natureweb.com)

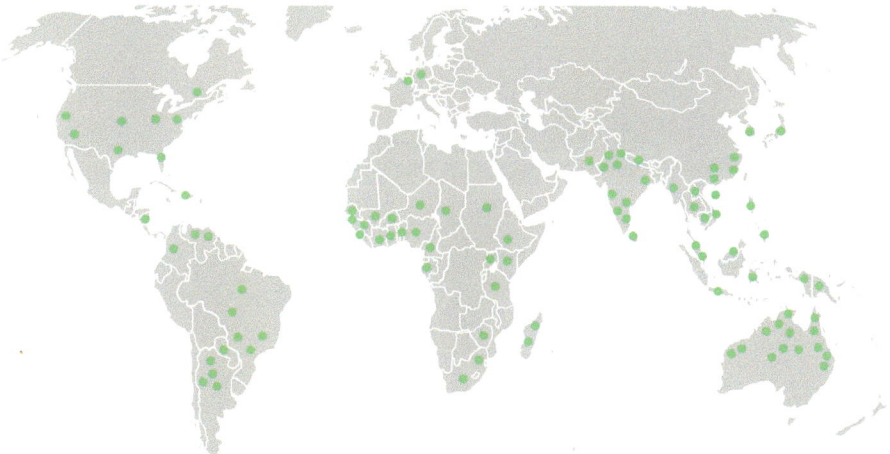

Fig. 18.4 Distribution area of the tailed cellar spider *Crossopriza lyoni* (green)

stage, during which they have just started with web building and prey capture, can catch a mosquito and suck it dry.

The tailed cellar spider probably originates from the area of Pakistan, Afghanistan, and Iran. Over the last 130 years (possibly even earlier, but there is no evidence), it has been introduced to America, Africa, Europe, as well as large parts of Asia and Australia (Fig. 18.4).

18.3   Marbled Cellar Spider *Holocnemus pluchei* (Cellar Spiders, Pholcidae)

The basic color of the forebody and legs is light beige to light yellow-brown. The forebody often has a dark band along the middle, and the sternum on the underside is black. Each leg is annulated brown-white-brown at two points, and in addition, they are patterned with many black dots or lines. The hind body is oval. The top side bears an irregular, brown-gray to reddish-brown marbled longitudinal band, which is usually framed in white. The sides of the hind body show white-silver wavy lines, and the underside has a wide, dark longitudinal band (Fig. 18.5). Both sexes are similarly colored. The body length of the females is 5–8 mm, and that of the males is 5–7 mm.

In the wild, males are mature from March to September, and females can be found all year round. These animals occur in their original area in the Mediterranean in dry terrain under stone blocks and in caves. Here, they build coarse-meshed webs with a woven-in **dome-shaped catching sheet**. When threatened, they show the typical trembling behavior of many cellar spiders, but they do not whirl around like the long-bodied cellar spider *Pholcus phalangioides*. Instead, they lower and raise their bodies rather quickly ("hopping"). The effect is the same: the contours of the body blur in front of the eyes of a predator such as a bird.

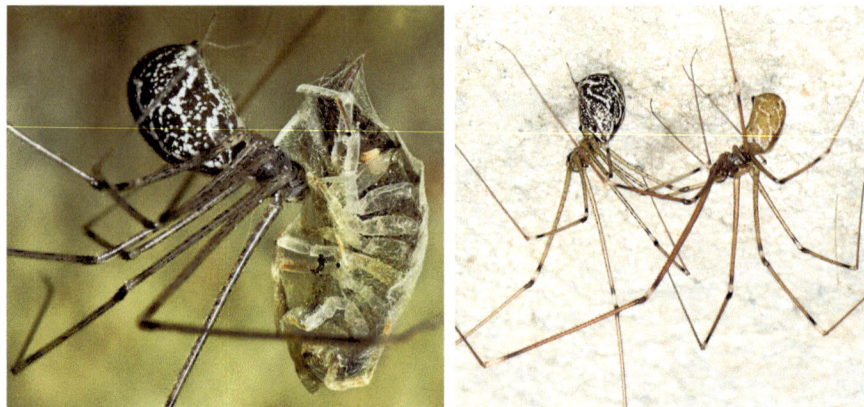

Fig. 18.5 Female of the marbled cellar spider *Holocnemus pluchei* sucking a woodlouse dry (**left**), and female and male (**right**) during courtship. (*Photos* left Dragiša Savić, right Jean-Philippe Taberlet)

Holocnemus pluchei tends to occur in **groups**, with webs built close to each other in places where a lot of food can be expected. If a web becomes heavily soiled or damaged, the spider often migrates to another free web instead of building a new one. There are no social interactions between neighbors. Each catches its own prey and threatens its neighbors if they come too close.

As is common with cellar spiders, females hold the loosely spun **eggs** with their mouthparts to guard them. However, unlike other species, *Holocnemus pluchei* does not seem to eat during this time. The female retreats into a protective spherical web of about 5 cm in diameter, into which she spins herself while guarding the eggs. An egg package contains about 50 eggs, and a female can produce eggs two or even three times. Their life expectancy can be about a year.

The step from the cave to the cellar is not a big one; thus, the spiders were probably able to penetrate into human buildings in their original Mediterranean area, where they also inhabit wall niches, sheds, and all kinds of outbuildings. Popular habitats include roof overhangs and attics. These spiders also like to sit under manhole covers or, depending on the climate, colonize the outsides of buildings.

In some situations, groups of several hundred spiders can come together, and through their spinning behavior, facades, windows, and parts of rooms become badly soiled. As you can imagine, this not only causes an aesthetic but also cleanliness and even hygiene problems. In California, *Holocnemus pluchei* was therefore designated an *urban pest*, which can perhaps be euphemistically described as an "urban problem."

Originally from the Mediterranean and Middle East, *Holocnemus pluchei* has shown tendencies to spread within and outside Europe since the 1960s. Today, this species is therefore found everywhere in Europe, except in Scandinavia and the Baltic States. In addition, it was introduced to the United States and Argentina in the 1950s, to Australia in the 1960s, and to Japan in the 2000s (Fig. 18.6).

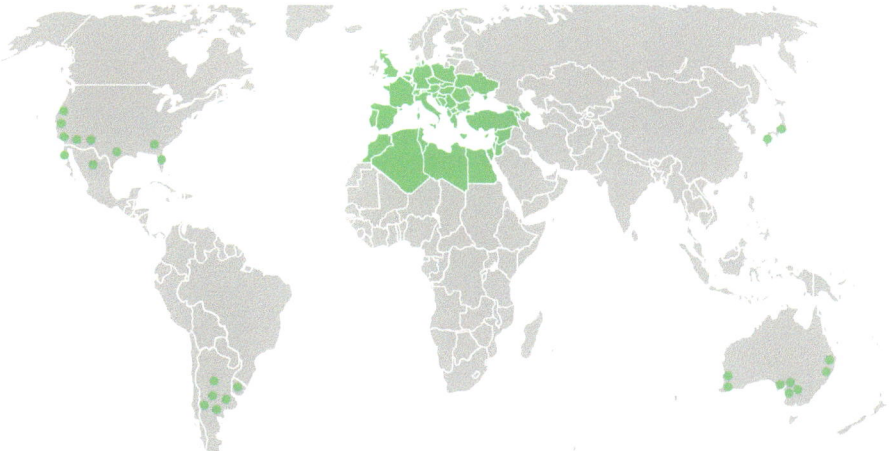

Fig. 18.6 Distribution area of the marbled cellar spider *Holocnemus pluchei* (green)

18.4   The Long-Bodied Cellar Spider *Pholcus phalangioides* (Cellar Spiders, Pholcidae)

Probably everyone knows them, but probably not everyone has noticed them: the long-bodied cellar spider (*Pholcus phalangioides*). This species originally lived in western Asia, probably in rocky, stony habitats and in the vicinity of cave entrances. The transition from such places to undisturbed, human-constructed spaces such as warehouses is not surprising. These habitats are similar, as is their microclimate, especially the very low humidity in the room. And since this spider occurs in storage rooms, it is not astonishing that it was also transported with trade goods. Thus, the first reports from Central Europe almost exclusively came from wine cellars and buildings where trade goods were stored, and that was over 300 years ago.

The long-bodied cellar spider can be distinguished from other cellar spiders by its size, the eight eyes (Fig. 4.2), the cylindrical hind body, and the generally pale, yellowish to light brown coloration. On the back of the forebody, there is a brownish pattern, which can be small and somewhat washed out. The brownish pattern on the hind body can be expressed as a more or less distinct longitudinal band or also as paired spots (Fig. 18.7). The body length of both sexes is 7–10 mm. The legs can be up to about 40 mm long. At the joints of the legs, black-white annulations or spots are often present. Despite their long legs, these animals appear remarkably inconspicuous. Cellar spiders occur in the house all year round and can have offspring at any time of the year. In captivity, they can live up to three years.

The long-bodied cellar spider is now found in every country in Europe, in every town, and probably also in every building. We assume that it has become one of the **most common spider species** in Europe. However, trade does not stop at the borders of Europe and has continued to increase under the trend of globalization. So, it is not surprising that the long-bodied cellar spider could spread with trade routes to

Fig. 18.7 Female of the long-bodied cellar spider *Pholcus phalangioides* in ambush (**a**), with an egg sac in the mouthparts (**b**), and with the freshly hatched young (**c**). (*Photos* (**a**) Eckhart Derschmidt, (**b**) Jörg Pageler, (**c**) Gordon Ackermann)

other continents too, and today it is found in many regions of North and South America, Africa, Asia, and Australia (Fig. 18.8). The distribution given here should always be understood as a minimum distribution because *Pholcus phalangioides* probably occurs in many other areas. Whether it is the most common spider species worldwide would be difficult to determine. But it is definitely the most widely distributed and most common house spider.

As a cellar spider, *Pholcus phalangioides* has extremely long, thin legs compared to its body, such that it seems to hover freely in its web, in corners of garages, cellars, attics, and also everywhere else in the house. It moves skillfully toward the finest vibrations, which are caused by insects or other prey, and carefully wraps the prey items in spider silk. This process is easy to observe, and it is obvious that it is important to the spider to keep its still-living and potentially dangerous prey at a distance using its long legs.

Once the prey has been sufficiently wrapped, the spider carefully looks for a suitable spot for the venomous bite, which it usually finds on or near the head. It injects its venom and waits for a while until the movements of the prey stop. Then feeding and digestion begin. Spiders digest their prey in front of the mouth, meaning they regurgitate digestive juice onto the prey, dissolve the prey over the course of several hours by pumping digestive enzymes back and forth, and finally suck it dry out of its chitinous shell. This requires tremendous muscle forces for the pharynx and the pumping stomach, which enable a cellar spider to, for example, suck a long-legged insect completely dry down to the tips of its toes.

The broad **prey spectrum** is remarkable and includes large and aggressive insects. This can be easily observed and is also an extremely interesting exercise for the spider fan, because in the web and especially on the ground underneath, the still-carefully wrapped remains of the sucked-out prey can be found (Fig. 2.2). In this way, you can easily collect the prey and thereby study what was on the menu in the previous weeks.

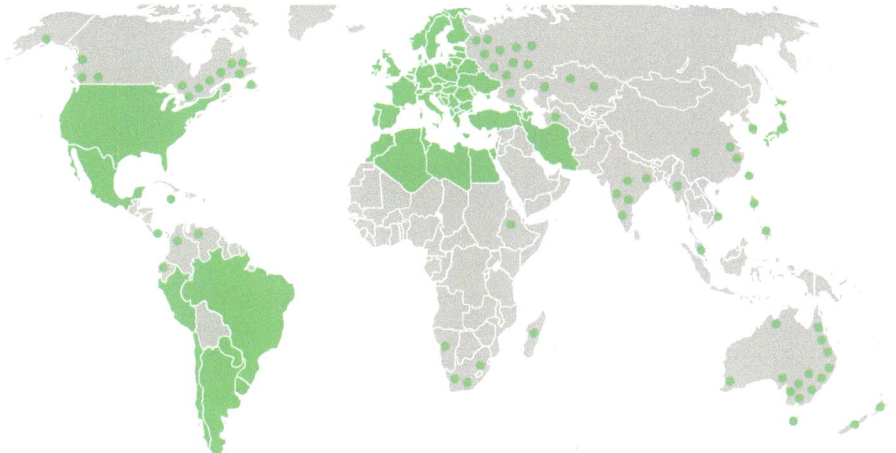

Fig. 18.8 Distribution area of the long-bodied cellar spider *Pholcus phalangioides* (green)

An examination over two years of the sucked-out prey remains in a toilet with an often open window in Germany showed that 38 spider species, including even the giant house spider *Eratigena atrica*, as well as 29 species of insects and other arthropods were preyed upon. In buildings where cellar spiders spread, most other spiders therefore disappear over time, and the number of insects decreases.

Cellar spiders can be observed very well, and therefore we recommend them as pets if you want to get more involved with spiders. In a **terrarium**, you can watch them build their web, overpower different prey, and suck them dry. By disturbing the spider, the trembling behavior can be triggered, in which it and surrounding web parts vibrate, such that the spider blurs in front of its enemies' eyes and is thus protected.

Their **brood care** is also an exciting experience. About two weeks after mating, the female lays one to two dozen eggs and holds them together with silk threads so that they look like a plump blackberry. She then carries this egg package around in her jaws for about two to three weeks until the spiderlings hatch. If a prey item becomes entangled in her web during this time, she can place the egg package in her web, overpower the prey and eat it, then pick up and guard the eggs again. The spider behaves similarly when she wants to clean or repair her web.

Even after hatching, the spiderlings remain with their mother as a ball of small dots and many long legs (Fig. 18.7c). Until the first molt, they do not eat any food. After molting, they sometimes share prey, or the mother allows some spiderlings to suck on the next prey item that she catches. Since cannibalistic feelings can also arise at this time, such spiderling groups then disperse rather quickly.

18.5   American Cellar Spider *Psilochorus simoni* (Cellar Spiders, Pholcidae)

The forebody is yellowish to light brown, slightly darkened in the middle, and the legs are also yellowish to light brown. The hind body is strongly spherical to oval, gray with a bluish hue, and the spinnerets are shifted forward (Fig. 18.9). The body length of females and males is only 2–3 mm, and the length of the front legs is about 10 mm.

The American cellar spider builds an **umbrella-shaped web** and hangs on the underside of the dome. It catches a wide range of insects with its web, including beetles and moths. However, when threatened, it does not show the typical trembling behavior of other pholcids.

Psilochorus simoni becomes mature after about half a year and occurs from February to August as an adult (mature) animal. After copulation, the female can produce at least four **egg packages**, each with 20–30 eggs, at appropriate intervals. The life expectancy can (at least in captivity) exceed one year.

Originally, this spider only occurred in a limited area in North America (California) in dark and cool places, such as caves, and was considered rare. In the meantime, it has also discovered cellars, especially wine cellars, as a habitat. Today, it can be found in all kinds of cellars, storage rooms, and other buildings. The cool

Fig. 18.9 Female (**a**), female with egg package (**b**), and male (**c**) of the American cellar spider *Psilochorus simoni*. (*Photos* Arnaud Henrard)

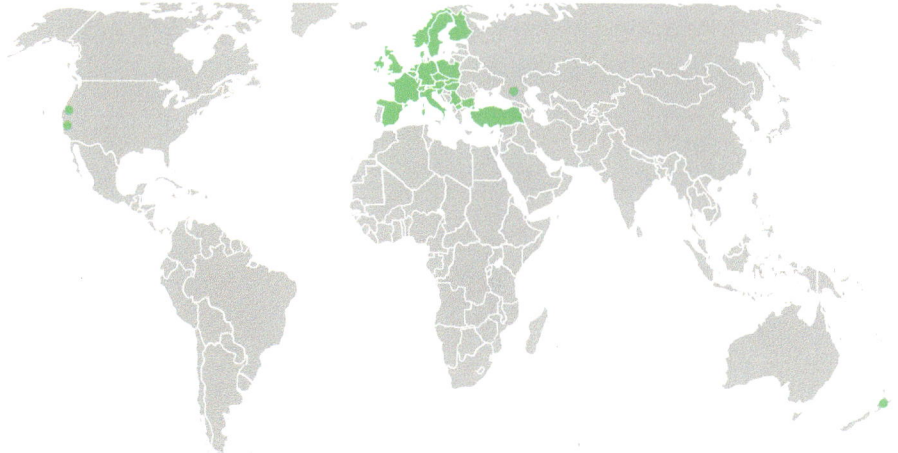

Fig. 18.10   Distribution area of the American cellar spider *Psilochorus simoni* (green)

temperature or special humidity conditions do not seem to be important parameters, but the habitat should be dark. The later so-called American cellar spider has since been introduced into Europe, Turkey, and New Zealand over the course of about 100 years (Fig. 18.10). Also in these areas, the spider does not seem to be common or is difficult to discover, as it prefers dark corners near the ground.

18.6   *Smeringopus pallidus* (Cellar Spiders, Pholcidae)

Both sexes of *Smeringopus pallidus* are similarly colored. The forebody is yellow-ish to light brown with a brown stripe in the middle. The legs are also light yellow to light brown and show a dark annulation at two points. The cylindrically elongated hind body is similarly light with a dark brown, sometimes purple-looking, pattern of paired spots on the back and underside (Fig. 18.11). The body length of the females is 6–7 mm, and that of the males is 5–6 mm.

This species originates from tropical Africa, although little is known about it from there. It probably lives in shady and moist places such as under stones and roots, in holes, and small caves. Here, the spider builds its domed **sheet web**, in which it sits at the highest point. When threatened, it shows the typical trembling behavior of cellar spiders.

With the usual goods transport pathways, *Smeringopus pallidus* was introduced to Central and South America and also to Asia and Australia. At least once it was (unknowingly) transported to Europe, but this did not lead to any permanent settlement (Fig. 18.12). This spider is found in basements, in all kinds of storage places, in garages, and also on the exterior walls of buildings.

18.7   *Spermophora senoculata* (Cellar Spiders, Pholcidae)

This species has only six eyes (Fig. 4.2) and is characterized by its small body size. Both sexes reach about 2–3 mm in body length and are similarly inconspicuously colored. The forebody and the legs are almost colorless to whitish, with a brownish

Fig. 18.11 *Smeringopus pallidus:* **left**, the smaller male and a female during courtship; **right**, a male viewed from above. (*Photos* Bernhard Huber)

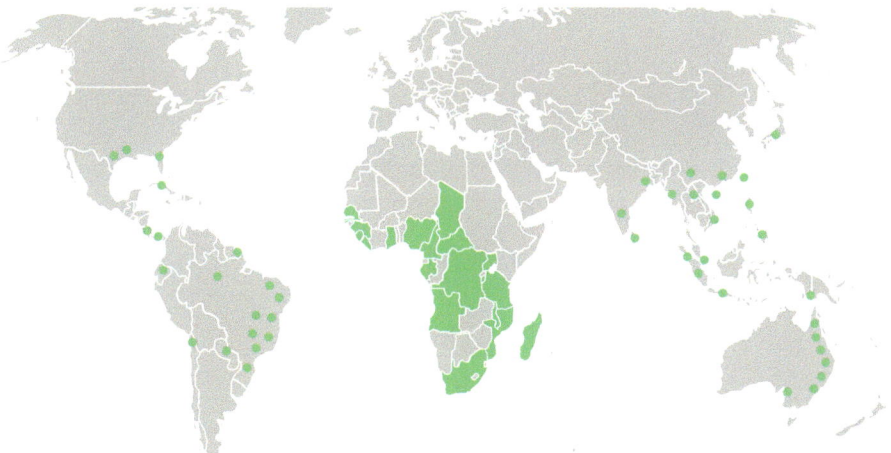

Fig. 18.12   Distribution area of *Smeringopus pallidus* (green)

spot on top of the forebody. The hind body is spherical and also very lightly colored, with three pairs of dark spots, some of which can be very pale (Fig. 18.13).

Spermophora senoculata builds remarkably small **space webs**, usually only 5–10 cm large, under whose capture sheet it lurks. It catches many small insects, including mosquitoes and moths. Unlike other pholcids, prey items are only slightly wrapped, and the spider tries to apply its venomous bite as quickly as possible. The webs of different individuals are built at a distance of 20–30 cm from each other, so that a sufficiently large safety distance to the next spider exists. It has been observed that these animals can be extremely cannibalistic. When threatened, the animals can show the typical trembling behavior of cellar spiders, but often they also try to quickly flee from their web.

From an egg package, 15–20 young spiders hatch, which in Europe mature in May and live until about September. A female can produce several egg packages in succession. The entire lifespan is about a year.

In their original habitat in the Middle East, these spiders live under stones and in caves. Where they occur in human buildings, they prefer dry, warm, and shady places. According to one study, they show a certain affinity for paper and are often found among books, folders, and boxes, but also under or behind furniture. In basement rooms, they are usually found in the upper area of the walls under the ceiling and on the ground near the floor. So far, this species has been introduced to the United States and the Caribbean, has spread to Europe, and is now also found in several Asian countries (Fig. 18.14). Interestingly, *Spermophora senoculata* seems to be spreading back into natural habitats in both the United States and southern Europe.

Fig. 18.13 Female of *Spermophora senoculata* with eggs. (*Photos* top Gordon Ackermann, bottom John Maxwell)

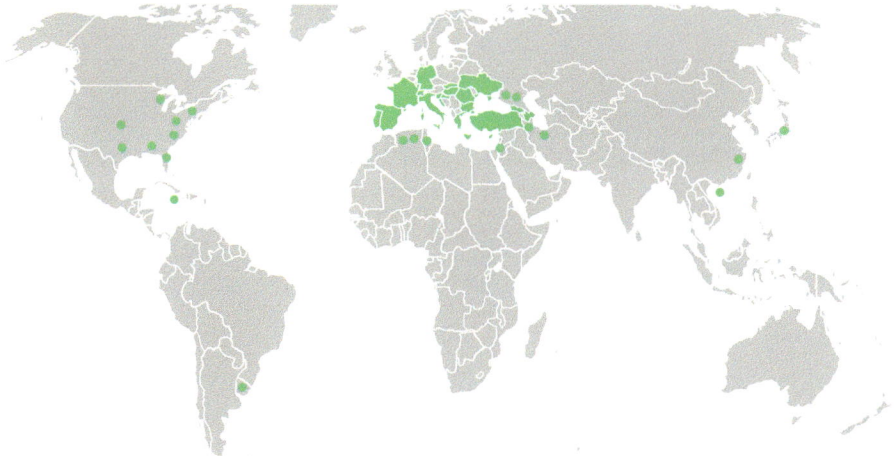

Fig. 18.14 Distribution area of *Spermophora senoculata* (green)

Nearly 13% of all spider species are jumping spiders (Salticidae). With over 6500 species, they are the most species-rich family of all. Jumping spiders are found throughout the world and are abundant in all terrestrial habitats. They do not build capture webs but stalk their prey before jumping on it to overpower it (hence their name). This is followed by a bite with the mouthparts, preferably in the head or thorax area, where the large muscle packages and nerve centers of their prey are located and where their venom works particularly quickly.

Spider silk is used by this family primarily for three purposes: to build a retreat, to protect the eggs in a silken eggsac, and as a **safety line** in case the jump should ever miss the landing site. The latter can be easily observed when you scare away a jumping spider on a tabletop with your finger and let it hop over the edge of the table: about 20–30 cm below the tabletop, it will hang in the air and shortly thereafter climb back up the safety line.

Jumping spiders have adapted to this hunting lifestyle through their entire body structure. The front middle eyes are particularly large and allow the animals to see sharply at a considerable distance. Optically, they are equivalent to telephoto lenses. In addition, these eyes are independently movable and have very good spatial resolution. The relatively small field of view (10°) is compensated for by a unique retina that can be moved very precisely in all three spatial directions using three pairs of muscles. This results in a field of view of almost 60°. Since the eight eyes are almost circularly arranged around the forebody (Fig. 4.4) and their fields of view overlap, they have spatial vision that allows them, for example, to estimate distances. Unlike most other spiders, jumping spiders are therefore diurnal, i.e., active during the day. Another special feature: all jumping spiders studied so far have **color vision,** which includes four colors; in other words, they can perceive red, green, blue, and also ultraviolet. For us humans, with our system based on 'only' three colors, it is hard to imagine what the colorful world of jumping spiders looks like.

Other adaptations of jumping spiders include a fairly compact body and short, sturdy legs. These bear **adhesive hairs** (scopulae), which allow them to move even

© Association for the Promotion of Spider Research 2024
W. Nentwig et al., *House Spiders - Worldwide*,
https://doi.org/10.1007/978-3-031-70448-2_19

on smooth surfaces. These can be leaves coated with wax, or in human settlements, smoothly painted window frames and doors or panes of glass.

The mating behavior of jumping spiders is adapted to their good visual sense. Jumping spiders are often strikingly colorful, with males usually more colorful than females, such that you can distinguish the sexes by their appearance alone. In addition, the males of many species perform a courtship dance in which the contrasting legs, pedipalps, and eye areas are important. Females often appear in gray and brown tones, making them better camouflaged. This is particularly noticeable in jumping spiders that are adapted to bark as their original habitat. To stand out during courtship, the adult males deviate from this camouflaged bark color and display, at least seen from the front, striking color patterns.

Here, we present the nine most common jumping spiders that we found in our study in buildings. For a key to these species, we would have to refer to illustrations of genital structures, which we want to avoid here. Instead, we invite you to compare a spider to be identified with the photos shown here.

19.1   Adanson's House Jumper *Hasarius adansoni* (Jumping Spiders, Salticidae)

With a body length of almost 9 mm, females of Adanson's house jumper, *Hasarius adansoni*, sometimes also called the greenhouse jumping spider, are quite noticeable. Males are slightly smaller but more contrastingly colored. In greenhouses and similar warm buildings, these animals are therefore hard to miss.

The forebody of the female is reddish-brown to brown-black. The eye area is surrounded by white or yellowish, sometimes almost orange hairs, which continue over the palps, which are impressively white or yellowish. The hind body is also surrounded by white-yellowish hairs and appears quite colorful, with a white-yellow stripe in the middle and orange stripes next to it. In some animals, however, the hind body is quite uniformly yellow-brown with washed-out reddish spots and dots. Males are darker but very noticeable due to their white-striped palps, a white U-shaped line on the forebody (open at the front), and one on the hind body (open at the back). The eye area is underlain with rust-red, and the hind body has a rust-red band in the middle, which can sometimes be very wide and is framed laterally by two or four white spots (Fig. 19.1). The coloration and markings can be quite variable.

Hasarius adansoni probably originated from the tropics of Africa or Asia but was then spread throughout the world with plants. As a tropical species, this spider cannot live in the wild in other climate zones, such as in Europe, but is limited to greenhouses and the like, where adult animals can be found all year round. In warm regions, *Hasarius adansoni* has also established itself in the wild, but these locations are often near **greenhouses**, indicating how this spider was able to spread (Fig. 19.2).

The greenhouse jumping spider is often found climbing on plants, walls, and rocks and then likes to hide under bark or on plant tendrils in a silken retreat. In

Fig. 19.1 Adanson's house jumper *Hasarius adansoni*: **left**, a female; **right**, a male. (*Photos* Michael Schäfer)

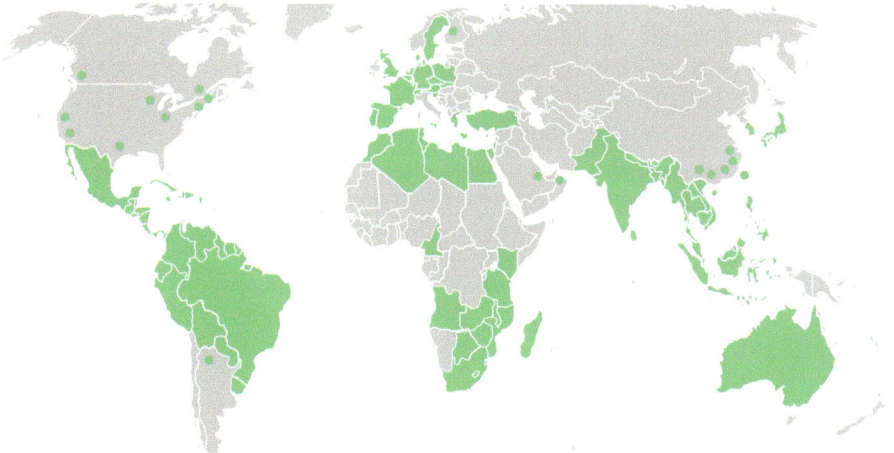

Fig. 19.2 Distribution area of Adanson's house jumper *Hasarius adansoni* (green)

warm countries, it can also be found outside of greenhouses on the interior and exterior walls of buildings. Like all jumping spiders, it preys on all possible insects that it encounters in its habitat by sneaking up on them and grabbing them with a jump.

19.2   Bark Jumping Spider *Marpissa muscosa* (Jumping Spiders, Salticidae)

A very lively species, with a body length of 14 mm in females (males 8 mm), *Marpissa muscosa* is one of the largest jumping spiders in Europe. In recent decades, this formerly rare jumping spider has undergone a significant shift in habitat and has now become common, allowing anyone to observe it in their homes in Europe.

The forebody of the spider is gray-brown to dark brown in color and covered with light gray hairs. The hairs around the front eyes are orange to reddish-brown, which are much more noticeable in the female than in the male. The legs are speckled or annulated in color like the front body and lightly haired. The hind body is light brown, medium brown, and speckled reddish-brown, with light spots. This pattern can be consolidated into color bands (Fig. 19.3).

Originally, *Marpissa muscosa* was considered to be a spider that lives on the **bark** of trees and also likes to hide under loose pieces of bark, hence its somewhat flattened body shape. Its coloration is well suited for such a lifestyle on and under bark because the spider is very well camouflaged there and not easy to spot. However, at some point, the bark jumping spider managed to penetrate our settlements. In this secondary habitat, it lives on the outside of house walls, on fences, railings, patio furniture, and also inside the house on the walls. There, its color camouflage is no longer helpful, and we can easily discover this quite conspicuous spider due to its size. This is relatively straightforward for us, as *Marpissa* is curious and regularly roams around its hunting territory looking for prey.

The bark jumping spider has expanded its habitat and come to occupy a much larger living area with its colonization of human buildings. However, it has not made the leap to another continent and is therefore a native species everywhere it

Fig. 19.3 The bark jumping spider *Marpissa muscosa*: **left**, a female; **right**, a male. (*Photos* Michael Schäfer)

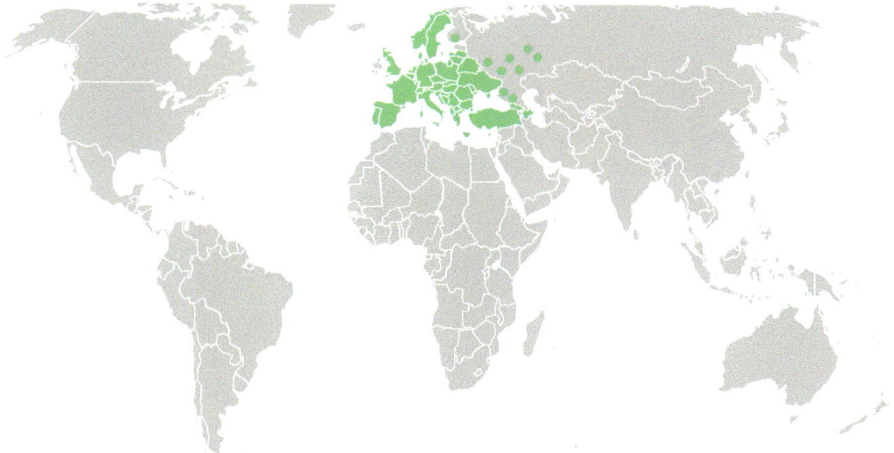

Fig. 19.4 Distribution area of the bark jumping spider *Marpissa muscosa* (green)

occurs today (Fig. 19.4). But who knows, maybe this next phase of expansion is yet to come?

The bark jumping spider eats almost everything in terms of small insects that appear in its environment and thus contributes significantly to keeping our houses insect-free. At night, it goes back into a retreat that it has set up in a crack in masonry or wood. Mating takes place in spring, followed by up to five eggsacs produced and guarded until the young spiders hatch. These spiders are mature at two years old.

Jumping spiders are completely harmless to humans. Due to their size and the curiosity of the animals, you can even play with them. If you hold out a finger toward them, they can climb up it and later jump onto a second finger that carefully approaches them. This game can be repeated several times with a little patience.

19.3 Gray Wall Jumping Spider *Menemerus bivittatus* and *Menemerus semilimbatus* (Jumping Spiders, Salticidae)

The basic color of these jumping spiders could be described as gray in gray, covered with grayish-white hairs. Together with shades of brown, gray is the optimal camouflage color when living on rocks and bark or colonizing the interior and exterior walls of our buildings.

The underlying color in both sexes is a sandy gray, with medium to dark brown or black-brown as the second contrast color. Females have a dark longitudinal band on the sides of the fore and hind body. The front eyes are surrounded by a yellowish or reddish band. In males, a similar longitudinal band is found on both sides of the forebody, but on the hind body, these continue as a middle band, so that the overall pattern resembles a 'Y'. In both sexes, the legs are annulated (Fig. 19.5). The color

Fig. 19.5 Female of the gray wall jumping spider *Menemerus bivittatus*. (*Photos* left Vida van der Walt, right Pierre Loria)

of the animals varies between individual specimens, from those that have a striking white-black contrast to others that appear generally very pale.

Menemerus bivittatus is a species of jumping spider that is relatively flat compared to many others. Originally living on rocks and tree trunks, it had no trouble venturing into human settlement areas where it could colonize many substitute habitats. Today, the gray wall jumping spider is mostly found on the exterior walls of buildings, but also occasionally inside houses. In addition, the animals can live on and in all sorts of structures built by humans in settlement areas, such as sheds or other extensions, gates, or fences. Females grow up to 9 mm long, while males reach up to 8 mm and are somewhat slimmer than the females.

The genus *Menemerus* currently comprises 63 species, almost all of which originate from the warm zones of Africa and Asia. Through human activity, *Menemerus bivittatus*, originally from tropical Africa, has spread throughout the world and can now be found on buildings on all continents (Fig. 19.6). Furthermore, at least seven other *Menemerus* species have been found repeatedly in and on buildings, and several species have also been spread to other continents.

In addition to *Menemerus bivittatus*, *Menemerus semilimbatus* also regularly appears on the interior and exterior walls of our houses. Originally a Mediterranean species, this jumping spider can now be found on several other continents (Fig. 19.6). It is noticeable for its white side stripes on the forebody and a dark brown area in between, in which there is a white triangle. The hind body is speckled in various shades of brown. Males also have striking black-and-white pedipalps (Fig. 19.7).

The **prey-catching behavior** of the jumping spider *Menemerus semilimbatus* has been studied in detail. Flies are common on walls, so it is not surprising that flies are caught primarily. Occasionally, other spiders and all sorts of small insects are preyed upon. Most insects are smaller than the jumping spider, and only in exceptional cases, which always appear spectacular to us, does it hunt larger insects. When observing *Menemerus semilimbatus* over a longer period of time, you will

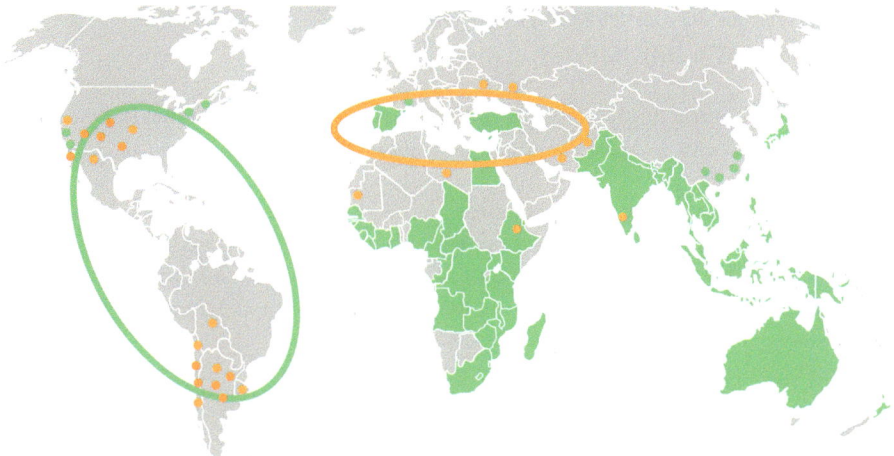

Fig. 19.6 Distribution areas of the gray wall jumping spider *Menemerus bivittatus* (green) and *Menemerus semilimbatus* (yellow-brown)

Fig. 19.7 Female (**left**) and male (**right**) of *Menemerus semilimbatus*. (*Photos* Michael Schäfer)

notice that the jumping spider prefers to approach flies from behind or from the side, as this reduces the risk of being detected by the fly. If the fly is looking directly at the spider, the predator makes a wide detour around the potential prey until it is beside or behind it. Then, the spider slowly creeps up on the insect. Once the distance is sufficiently small, *Menemerus semilimbatus* jumps powerfully onto the fly, holds it with its forelegs, bites into the area of the head or thorax, and injects its venom. An interesting side result of this study is that only about 8% of all observed jumping spiders successfully caught prey during the long observation period. This shows that it is not so easy to hunt successfully as a jumping spider.

19.4   The Bold Jumper *Phidippus audax* (Jumping Spiders, Salticidae)

The large and contrastingly colored bold jumper *Phidippus audax* is widely distributed throughout North America. It can be found in fields, prairies, yards, and open spaces on the edge of forests. This black jumping spider is noticeable due to its green shimmering mouthparts and three white or orange spots on the hind body (Fig. 19.8). Females are 8–18 mm long, while males are 6–15 mm. Like all jumping spiders, they have very acute vision and actively hunt during the day. Jumping spiders can jump several times their body length, and for large spiders like *Phidippus audax*, these are considerable distances.

Although the bold jumper *Phidippus audax* sometimes stays in houses, it is more often found in the vegetation around houses. These jumping spiders are indeed brave hunters. The spider does not hesitate to pounce on an insect that is as large as, or slightly larger than, itself. Normally, the spider does not use silk to secure its prey. Instead, it injects venom with its mouthparts as soon as possible and quickly begins to feed after catching an insect. However, the spider is always connected to the ground behind it by a safety thread.

Males and females build soft, thick silk hideouts under objects in the garden, such as under flower pots, wood, and bricks, and sometimes even in our letterboxes. Females build their eggsacs also within their silk-spun retreat. They are mainly active during the warmer months and overwinter in their silk hideouts, often under bark. During winter, you can see groups of these jumping spiders close together in particularly suitable places where they are well protected from frost.

Phidippus audax is found in the United States and adjacent areas of Canada and Mexico. Spiders of this species were, strangely enough, brought to Hawaii and the Azores and the Nicobar Islands, but not to other continents. Individual animals have

Fig. 19.8 Female (**left**) and male (with a captured spider) (**right**) of the bold jumper *Phidippus audax*. (*Photos* left Jim T. Johnson, right Ken Childs)

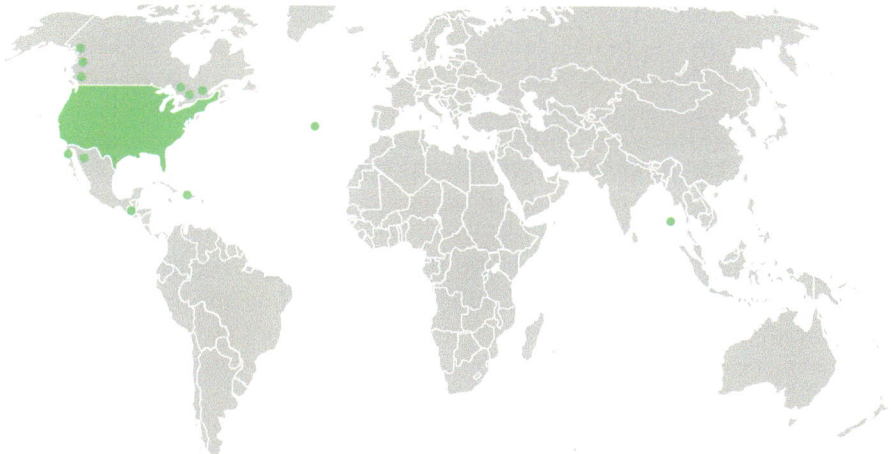

Fig. 19.9 Distribution area of the bold jumper *Phidippus audax* (green)

been found in Brazil and Holland, but have apparently not yet established themselves there (Fig. 19.9).

As with all other species of jumping spiders, *Phidippus audax* also uses visual signals during **mating**, and the male dances around his lover. If the female is mature and interested in mating with the male, she responds with less dramatic dance movements. The male then approaches the female and mates with her. Mating can apparently, depending on the temperament, last between 14 min and over 2 h!

The bold jumper *Phidippus audax* is not aggressive and, in fact, is quite good-natured. Therefore, it is also advertised as a suitable **terrarium spider** and is available in pet stores. However, it is a rather large spider and can, if it feels threatened, bite in defense. The bite is said to be painful, but the venom is harmless to humans; thus, the pain subsides after a few minutes.

19.5 The Pantropical Jumper *Plexippus paykulli* and *Plexippus petersi* (Jumping Spiders, Salticidae)

In many regions, it is often reported that people are happy when the pantropical jumper *Plexippus paykulli* is found in and on their houses. Females reach up to 12 mm in body length, and males up to 10 mm. So these are quite large jumping spiders that can hunt a wide range of prey, including many insects that are harmful or annoying to humans, especially mosquitoes and cockroaches.

Like all jumping spiders, *Plexippus paykulli* is **diurnal** and hides in a silken retreat at night. On buildings, however, it has apparently learned that there are many insects attracted by light near the **house lighting** and similar light sources at night. Therefore, you can regularly observe the jumping spider hunting prey at such light sources even at night.

Females are well camouflaged in coloration and therefore inconspicuous. Their basic color is light gray to light brown with medium brown markings. This results in a bright longitudinal stripe that extends from the forebody over the hind body and shows two or three bright transverse stripes or bright spots on the back. The eye area is set off in front with a reddish-brown color. On the legs, you can see some black bristles; otherwise, the body hair is white (Fig. 19.10).

Males are more contrasting and thus more conspicuously colored. The basic color of the body and legs is whitish-yellow with two dark brown to black longitudinal stripes on the fore and hind body. The front legs, in particular, are provided with brown-black dots and longitudinal stripes. Viewed from the front, whitish and reddish-brown vertical stripes alternate in the eye area (not visible in Fig. 19.10). Especially considering their large distribution area, it is not surprising that the coloration of *Plexippus paykulli* varies greatly across the world.

Males seek the proximity of females, and both can then be observed together in the female's retreat, where mating also takes place. For **egg laying**, the female builds a fairly large silken retreat with a diameter of 25–35 mm in a protected area, into which it lays 30–60 eggs. The eggs are guarded by the female for 3–4 weeks until the young hatch. In warmer areas, adult spiders can be found almost all year round.

Plexippus paykulli is a jumping spider from the tropical areas of Africa, which has now been spread throughout the world by humans. Apparently, it sought the proximity to human dwellings early on, such that it could also be easily spread with the goods that humans transport worldwide. Today the pantropical jumper can be found on all continents in warmer areas, equally inside buildings as well as outside on houses and outbuildings (Fig. 19.11).

Fig. 19.10 Female (**left**) and male (**right**) of the pantropical jumper *Plexippus paykulli*. (*Photos* Michael Schäfer)

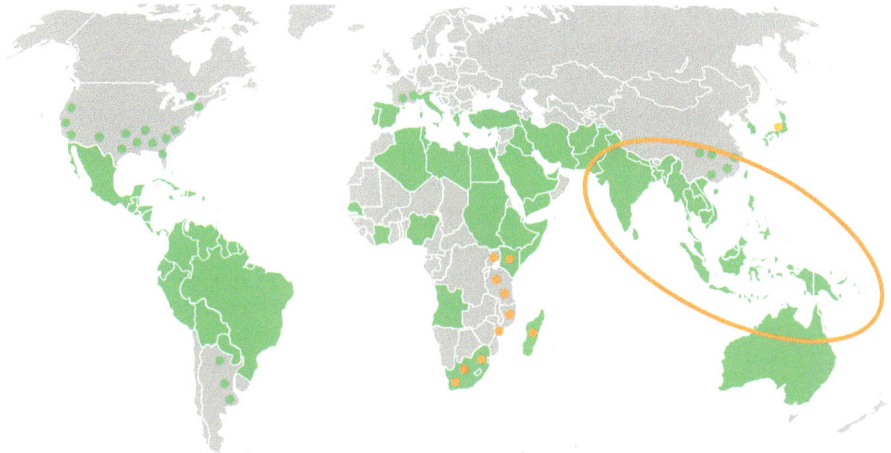

Fig. 19.11 Distribution areas of the pantropical jumper *Plexippus paykulli* (green) and of *Plexippus petersi* (yellow-brown)

Fig. 19.12 Female (**left**) and male (**right**) of *Plexippus petersi*. (*Photos* Joseph K. H. Koh)

Within the genus *Plexippus*, there are several difficult to distinguish species, some of which can also be found on buildings. In Asia and Africa, this particularly applies to *Plexippus petersi*, which can even be more common than *Plexippus paykulli* in Asia (Fig. 19.11). *Plexippus petersi* is characterized mainly by the fact that in males, the dark stripes on the forebody are more separated and shorter; females are barely distinguishable based on external features (Fig. 19.12).

19.6   Woolly Wall Jumping Spider *Pseudeuophrys lanigera* (Jumping Spiders, Salticidae)

This small, rather cute-looking jumping spider (females are up to 6 mm long, males slightly smaller) originally comes from rocky habitats in southern Europe. A few decades ago, the woolly wall jumping spider made three significant changes within a short time with respect to its habitat. First, it discovered that many walls occur in human settlement areas, which opened up completely new possibilities for it as a **substitute habitat**. Second, it was able to significantly expand its distribution when it advanced into areas north of the Alps that it had not previously inhabited. Third, it managed to colonize North America with a huge leap across the Atlantic (Fig. 19.14). In light of this dynamism, it can be assumed that *Pseudeuophrys lanigera* will continue to spread strongly in the coming decades.

The underlying color of this jumping spider is whitish yellow to medium brown. The sides of the fore and hind body are brown or brown-black spotted, such that the upper part of the forebody and the middle of the hind body have a light band. The eyes are rimmed yellow to reddish brown, the legs are annulated brown, and the palps are white-haired. Males are similar in color to the females but often appear darker (Fig. 19.13).

Pseudeuophrys lanigera can also be found on the outer walls of buildings but is somewhat more common inside buildings. There, the jumping spider can occur all year round as an adult animal. It preys on all kinds of small insects, such as psocids (bark or book lice), which are common in many houses. If conditions are appropriate for it, this spider can be quite common.

Fig. 19.13 Female (**left**) and male (**right**) of the woolly wall jumping spider *Pseudeuophrys lanigera*. (*Photos* Michael Schäfer)

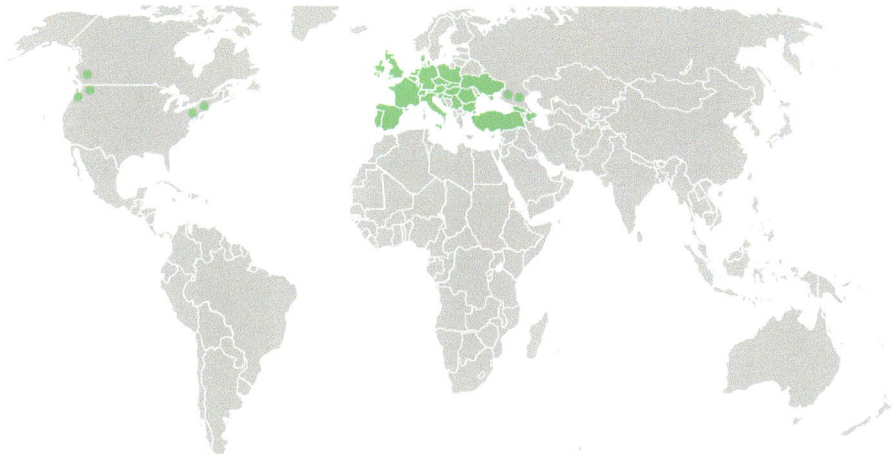

Fig. 19.14 Distribution area of the woolly wall jumping spider *Pseudeuophrys lanigera* (green)

19.7   Zebra Jumping Spider *Salticus scenicus* (Jumping Spiders, Salticidae)

The zebra jumping spider, *Salticus scenicus*, is unmistakable due to its black and white pattern, even though there are a total of 45 species in the genus *Salticus*, many of which look very similar. The basic color of their body is dark, predominantly black. There are three stripes of white hairs on the hind body, with the one at the front edge being continuous and the two rear stripes being interrupted in the middle. Often, the midline of the hind body is brown; this is more pronounced in females than in males, but the brown color can also be completely absent. The forebody is essentially black but can also have brown spots and usually has two pairs of white spots, which can be so large that they appear as a large X. The legs are patterned brown-black with strong white hairs, which are less pronounced at the leg joints. Males are colored the same as the females (Fig. 19.15). However, they have noticeably long mouthparts, which they use to impress females during courtship or to deter other males during combat.

With a body length of 5–7 mm for females and 5–6 mm for males, the zebra jumping spider is among the smaller jumping spiders. Like all jumping spiders, it has relatively short but strong legs compared to its body. It is a diurnal hunter and a particularly sun-loving species, which prefers to stay on south-facing house walls, window sills, or fences in the vicinity of settlements. At night and during bad weather, it hides in its **retreat**, which can be concealed in a wall crack, a pile of wood, or even in a letterbox. Mating also takes place within the retreat; thus, males and females can occasionally be found living together peacefully in such cohabitation retreats. The eggsac, which contains an average of about 30 eggs, is also laid within the retreat.

Fig. 19.15 Female (**left**) and male (**right**) of the zebra jumping spider *Salticus scenicus*. (*Photos* Michael Schäfer)

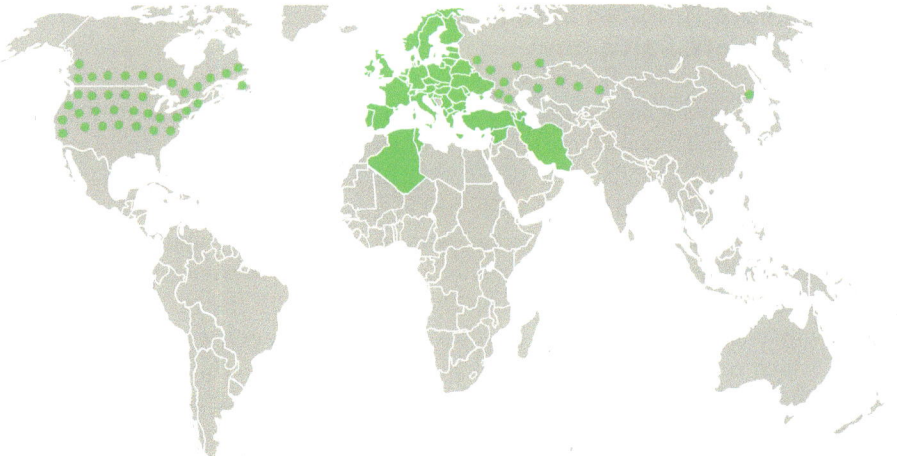

Fig. 19.16 Distribution area of the zebra jumping spider *Salticus scenicus* (green)

Although *Salticus scenicus* is not very large, the huge front middle eyes are very recognizable, giving the jumping spider a "face," so to speak, for us humans. With these large middle eyes, it not only targets prey but also a human finger in front of it or the lens of a camera. It usually moves very agilely, raises its forebody, runs sideways or even backwards, and may even jump onto your finger. Perhaps that's why the zebra jumping spider, unlike some other spider species, is perceived by most people as a very cute little creature.

Salticus scenicus is found all over Europe, into Russia, the Caucasus, Kazakhstan, and Iran, but has also spread to North Africa as well as the United States and Canada (Fig. 19.16).

Spitting Spiders (Scytodidae)

One could argue that spitting spiders are a rather insignificant family. Currently, only about 240 species are known, which, however, occur throughout the world. They are small to medium-sized and have six eyes, which are arranged in three groups (Fig. 4.3). Scytodids stand out due to their long, thin legs and the noticeably large forebody. This is as large as the hind body and, quite unusual for spiders, sometimes even slightly larger. Hidden inside is a **glue gland**, which enables spitting spiders to use a unique method of prey capture, probably unparalleled in the rest of the animal kingdom. Most scytodids do not build webs but roam slowly at night in search of prey. Only some tropical species build irregular webs, in which they lie in wait for prey.

Spitting spiders, as the name suggests, have developed one of the most spectacular and, for many people, completely unexpected hunting methods: they "spit" **glue** at their prey across two zigzag pathways at the same time, such that the victim becomes stuck to the ground and is therefore completely defenseless (Fig. 20.1). However, the term "spitting" is not quite correct, as the glue does not come from the mouth but is thrown out of the fangs, which are particularly short and can thus be precisely aimed at the target. The original **venom gland** produces the individual components of the venom in the front part, while in the greatly enlarged rear part, it produces glue and silk threads.

With their "glue cannon," a 5 mm long spitting spider can shoot at a distance of 20–30 mm and does so extremely fast, in about 2–3 hundredths of a second, such that we humans cannot see it. What the spitting spider shoots at its prey can best be understood as a composite material, in which a tough adhesive glue is additionally stabilized by embedded short silk-like strands. This material is not identical to the silk produced by the silk glands in the hind body. The legs, wings, and entire body of the prey animal are glued to the ground, and it can no longer escape. Presumably, the glue does not contain toxic substances, which were previously assumed to diffuse into the prey animal. Instead, we see that the spider, after having successfully glued down its prey, proceeds to give it a venomous bite.

© Association for the Promotion of Spider Research 2024
W. Nentwig et al., *House Spiders - Worldwide*,
https://doi.org/10.1007/978-3-031-70448-2_20

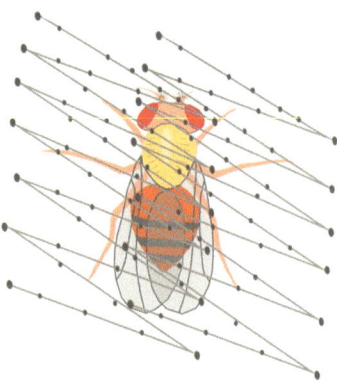

Fig. 20.1 Fruit fly glued by the common spitting spider *Scytodes thoracica* (**left**) and schematic "glue restraint" (**right**). (*Photo* Jutta Ansorg, *Figure* Wolfgang Nentwig)

Several *Scytodes* species have quickly realized that life inside human buildings can be quite comfortable. It is therefore not surprising that some were transported across the world with human goods and were able to settle in many countries, regularly occurring in buildings. The most frequent example is the common spitting spider, *Scytodes thoracica*, which will be described in more detail below. In North and South America, Africa, and from India to Japan, *Scytodes fusca* can be found in houses. In South America, *Scytodes globula* can also be encountered. In Central and South America, as well as from Southern Europe to Inner Asia and India, we can find *Scytodes univittata*.

20.1   The Common Spitting Spider *Scytodes thoracica* (Spitting Spiders, Scytodidae)

"Small but mighty" could be said about the common spitting spider, *Scytodes thoracica*. It is one of the most common house spiders, and once you recognize it, you will find it again and again. Nevertheless, *Scytodes thoracica* is hardly noticeable. For one, it is nocturnal and very shy. As soon as a light comes on somewhere or it senses the slightest vibration, it retreats back into a hiding place. On the other hand, with a body length of 4–6 mm, it is a rather small spider. Males are on average smaller than females, but only insignificantly. The fore and hind bodies are similarly colored, namely sandy with brown-black spots. The legs are also sandy, slender, and darkly annulated (Fig. 20.2). With this coloring, it is very well camouflaged, especially on sandy ground or on light natural stones.

Its inconspicuousness is also because *Scytodes thoracica* never runs fast. When it is hunting, it even moves very slowly. One could almost say it walks elegantly. It can also run more briskly, for example, when danger approaches, but a quick scurry, which many people often find frightening in spiders, is never seen. It remains to be determined whether it cannot run fast or simply does not want to run fast.

Fig. 20.2 A female of the common spitting spider *Scytodes thoracica* (**left**) and with eggsac (**right**). (*Photos*: left Jutta Ansorg, right Jean-Philippe Taberlet)

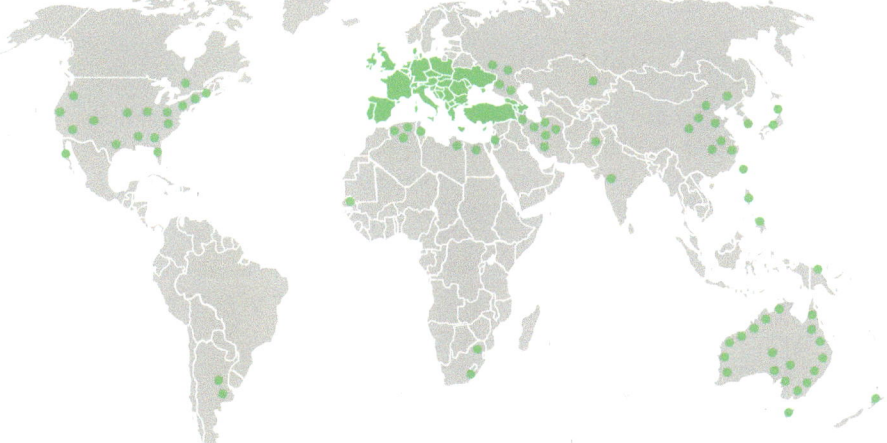

Fig. 20.3 Distribution area of the common spitting spider *Scytodes thoracica* (green)

One might think that the glue-slinging mechanism only works with very small animals and that spitting spiders are therefore severely limited in their choice of prey. However, this is not the case, because prey animals that are the same size as the spider or even a little larger are gladly accepted. It seems only to be important that the insects are thin-skinned, as the glue technique does not work so well against armored insects like beetles and ants. The prey of the common spitting spider therefore consists mainly of flies and mosquitoes, silverfish, small moths, and spiders. This is surprising at first glance, but *Scytodes thoracica* can successfully overpower other spiders, as they cannot counter the glue-slinging technique.

Originally, *Scytodes thoracica* comes from the Mediterranean region. There, it lives under stones and in rock crevices. Now, it is almost spread across the world (Fig. 20.3) due to human activities, but in cooler regions, it only occurs in houses.

Fig. 20.4 Courtship in the common spitting spider *Scytodes thoracica*: the female is on the left, and the male is on the right. (*Photo* Gordon Ackermann)

There, it likes to hide during the day behind pictures, skirting boards, shelves, and the like.

Mating in *Scytodes thoracica* can take place throughout the year, so you can encounter both males and females all year round (Fig. 20.4). Females lay about 20–35 eggs, wrap them with a few threads into a loose eggsac, and then carry this under their forebody until the young hatch (Fig. 20.2).

Most spiders only live for about a year. However, in *Scytodes thoracica*, females can live up to 3 years and males up to 2 years. In captivity, some animals have even reached an age of 4 or 5 years.

The family of tube-dwelling spiders, Segestriidae, comprises around 180 species worldwide. They are predominantly small to medium-sized spiders that have only six eyes (which you might only see at a second glance, Fig. 4.9), have a cylindrical body structure, and betray themselves through a characteristic leg position. **Three pairs of legs** are directed forward, with only one pair toward the back (Fig. 21.1). Especially when the animals are in their retreat and waiting for passing prey, it is remarkable to see these six legs. With this lurking position, a prey animal can probably be caught faster or more securely than if only four legs were stretched forward. However, this would still need to be scientifically tested.

Segestriids live on rocks or tree bark and use cracks and holes to set up their tubular retreat, where they lurk and wait for passing prey. From the edge of the tube, individual threads lead into the surroundings, serving as tripwires or for transmitting vibrations to the tube. The tube-dwelling spider can then purposefully run toward a potential prey animal. From this original habitat, it was easy for segestriids to expand into human settlement areas, where they find ideal habitats on walls and wooden structures in buildings.

21.1　Green-Fanged Tube-Dwelling Spider *Segestria florentina* (Tube-Dwelling Spiders, Segestriidae)

It could come from a science fiction film because it is large, dark, and fast: the green-fanged tube-dwelling spider *Segestria florentina*. Females of this imposing species can grow up to 22 mm long, making it one of the largest spiders in Europe. Its cylindrical body is an adaptation to life in tubes. It is very darkly colored, so the pattern of its hind body is hard to recognize (Fig. 21.2). Also striking are its dark green, shimmering chelicerae (hence the name), with which it can exert a powerful bite. However, its bite is completely harmless to humans.

The green-fanged tube-dwelling spider is found around the Mediterranean, where it lives under stones and in rock walls, sometimes also under roots or leaves.

© Association for the Promotion of Spider Research 2024　　　165
W. Nentwig et al., *House Spiders - Worldwide*,
https://doi.org/10.1007/978-3-031-70448-2_21

Fig. 21.1 The green-fanged tube-dwelling spider *Segestria florentina*, **left** in a lurking position in front of its tube, and **right** in its web for prey capture made of individual silk threads. (*Photos* Gordon Ackermann)

Fig. 21.2 (**a**) The green-fanged tube-dwelling spider *Segestria florentina*, (**b**) the Bavarian tube-dwelling spider *Segestria bavarica*, (**c**) the common tube-dwelling spider *Segestria senoculata*. (*Photos* (**a**) Gordon Ackermann, (**b**) Dragiša Savić, (**c**) Pierre Oger

It also likes to inhabit the cracks of tree bark. Further north, it often adopts holes and cracks in walls, including house walls, as a **substitute habitat**. Toward the east, its natural distribution extends to the European part of Russia and to Georgia. It has also been introduced by humans to Brazil, Uruguay, and Argentina (Fig. 21.3).

Its web consists of a wide-open silk tube, which narrows toward the back. From the front edge of this tube, radial **signal threads** ("fishing lines") run along the

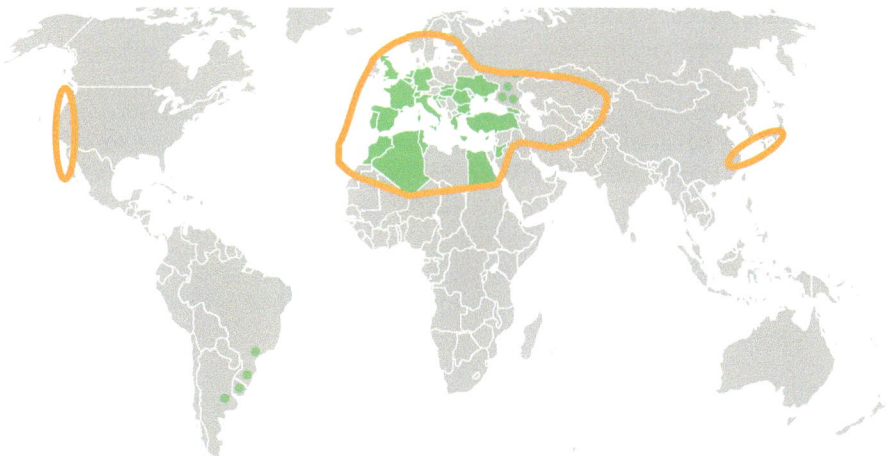

Fig. 21.3 Distribution area of the green-fanged tube-dwelling spider *Segestria florentina* (green) and the area where species of the genus *Segestria* are likely to occur (yellow-brown)

surface of the surrounding substrate. The spider sits in a lurking position at the entrance of its tube and has its front six legs stretched along six of the signal threads. The pedipalps also lie on the spider silk. This leg position of all *Segestria* species, in which the third pair of legs is directed forward instead of backward, is peculiar among spiders (Fig. 21.1). If a passing insect causes the signal threads to vibrate, the spider perceives this with the help of highly sensitive sensory organs on the legs and pedipalps. It immediately rushes out of the tube and attacks the prey with a venomous bite. The spider appears remarkably "self-confident" – the English spider researcher William Syer Bristowe tried to prevent a spider lured out of its dwelling tube from retreating using a pencil, whereupon it immediately bit fiercely into the pencil!

Segestria florentina has developed its own **hunting technique** for insects that can bite or sting powerfully, such as bees. These are bitten from behind, approximately in the middle of the body, and then the body of the prey is essentially folded in half. The spider then moves backward into the retreat with the prey, such that the mouthparts or the sting of the insect always point away from the spider. After a short time, the venom paralyzes the prey.

A mating-ready male first seeks the web of a female and shakes her signal threads to draw attention to himself. If the female remains calm, the male approaches, grasps her hind body with his mouthparts spread, and mates with her. The oval egg-sac contains up to 180 eggs.

About 30 years ago, it was reported that *Segestria florentina* often "sleeps" at night. During this resting behavior, it hangs completely limp, with the forebody directed upwards, seven legs in its tube, and one hind leg bent against the hind body. If disturbed, it wakes up instantly. At that time, nothing could be done with this observation, but today we know that a kind of **sleep behavior** is being found in more and more spider species. This consists of a rest or recovery phase with reduced reaction to environmental stimuli. Whether all spiders "sleep" in this sense, and how much they need it, cannot currently be estimated.

In Europe, two other species inhabit house walls: the Bavarian tube-dwelling spider *Segestria bavarica* and the common tube-dwelling spider *Segestria senoculata* (Fig. 21.2), sometimes called the snake-back spider because of its patterning. In North America and Asia, additional *Segestria* species are found (Fig. 21.3). The biology of these species is similar to that of the green-fanged tube-dwelling spider. Older records of tube-dwelling spiders, especially from tropical and subtropical areas, would still need to be checked – they could also belong to species from the closely related genus *Ariadna*.

Flatties (Selenopidae)

22

The family of flatties, or Selenopidae, is found throughout the world, especially in tropical, subtropical, and Mediterranean regions, but also in deserts and up to an altitude of 2500 m. Flatties do not build webs but are **nocturnal ambush predators**. These animals have a strongly flattened body and can hide well in narrow crevices and cracks. Tree bark and rocks are therefore their natural habitat. Their long legs are turned slightly outward, as in both crabs and crab spiders, allowing these spiders to move at lightning speed.

The small family of flatties currently comprises nine genera with 280 species, including over 130 species in the eponymous genus *Selenops*. These are all very similar in size and color and are therefore usually only distinguishable by their sexual organs and through examination under the microscope.

Presumably, like many other spiders, members of this family have traveled to other regions via human transport goods in ships and thus also reached temperate zones. Outdoors, they can be found in grass, bushes, under stones, under bark, and in tree stumps. In certain regions of Africa, they are also observed in agriculturally used areas such as avocado or macadamia plantations.

22.1 The Flatty Spider *Selenops radiatus* (Flatties, Selenopidae)

The common flatty spider *Selenops radiatus* appears as a spotted, brownish disk with strangely angled legs. This peculiarly flat spider grows up to 16–19 mm in body length. The forebody, as well as the hind body, is strongly flattened and varies in color from grayish or brownish to sometimes even slightly reddish, with black, brown, and gray lateral spots and patterns (Fig. 22.1). The hind body is round to oval and densely haired. The forebody is strikingly round with the eye region set forward. The eyes are arranged in two rows, six very far forward and two quite large ones on each side (Fig. 4.8).

© Association for the Promotion of Spider Research 2024
W. Nentwig et al., *House Spiders - Worldwide*,
https://doi.org/10.1007/978-3-031-70448-2_22

Fig. 22.1 Female of the flatty spider *Selenops radiatus* (**above**) in a building and male (**below**) on tree bark. (*Photos* Gordon Ackermann)

The long legs are usually brownish banded and adorned with striking white tufts of hair and spines. When the spiders are in their ambush position, waiting motionless for prey for a very long time, their legs are evenly arranged around the body, so that each covers an equally large area for detecting **vibrations**. Males and females are barely distinguishable, with females being slightly larger.

With its body speckled in shades of brown, the common flatty spider blends perfectly into the background. Thanks to the flat, disk-like shape of its body, it can hide in the narrowest cracks and crevices when in danger. This spider is incredibly fast; it can dart sideways at up to 63 body lengths per second. At this speed, it is one of the **fastest animals** known. And if that's still not fast enough to escape its enemies, the common flatty spider can shed individual legs, thus creating confusion.

Hopefully, it doesn't have to do this in our dwelling places, as we should be grateful for its presence. The common flatty spider hides in narrow gaps behind door and window frames, between roof beams and masonry, behind furniture and pictures, in fact, in any sort of small hiding place. It feeds on all manner of insects such as mosquitoes and fruit flies. But it is also said not to be averse to cockroaches, silverfish, and many other similar insects, thus freeing us from several pests.

The common flatty spider does not build webs but hunts freely, especially at night. Light attracts this spider, as if by magic, perhaps because it has learned that many insects buzz around lamps, which it can then easily prey upon. The common flatty spider can attack its prey not only from the front but also from the side. It does this with incredibly fast, abrupt movements of its legs. However, due to their nocturnal and very hidden lifestyle, most people will never realize that they have the flatty spider in their house.

Females attach their flat, round, and paper-like **eggsacs** outdoors under stones and in houses to wooden beams. Due to their whitish color, they are quite noticeable, especially since they can reach twice the body length of the females.

Selenops radiatus is a widely distributed species in Africa and occurs in the north up into southern European countries. In the east, the distribution area extends to South China and Sumatra (Fig. 22.2), although it is still not clear where, today, in Southeast Asia *Selenops radiatus* occurs and why there are so many gaps in its records from Africa. In the end, it is also conceivable that individual occurrences at the edges of its distribution area are due to unintentional spread by humans. The global distribution area of the many other very similar species in the genus *Selenops* is also shown in Fig. 22.2.

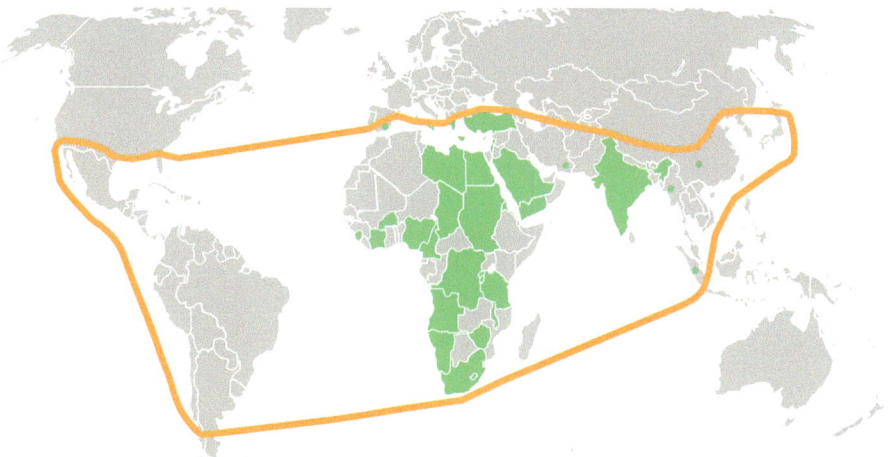

Fig. 22.2 Distribution area of the flatty spider *Selenops radiatus* (green) and the genus *Selenops* (yellow-brown)

Recluse Spiders (Sicariidae)

23

The small family includes about 200 species of six-eyed spiders. Included in this family is the genus *Loxosceles*, commonly referred to as recluse spiders or violin spiders (because of the marking on their forebody). These spiders are small to medium-sized, and their six eyes are arranged in three groups (Fig. 4.10). Recluse spiders naturally occur only in North and South America, in Africa up to the Mediterranean region including southern Europe, and sporadically in Asia. They are absent in Southeast Asia and Australia. These spiders do not build webs and live as ambush predators.

23.1 Recluse Spiders of the Genus *Loxosceles* (Recluse Spiders, Sicariidae)

Recluse spiders in the genus *Loxosceles* (Sicariidae), including the notorious North American brown recluse, are among the most feared spiders in the world as their bite can, in very rare cases, lead to death within 12–30 h. However, bites are extremely rare and usually harmless. Only in a few cases does a complex medical situation arise, called systemic loxoscelism, the cause of which is ultimately unknown but can have fatal consequences. *Loxosceles* bites rarely cause such severe reactions (an estimated few deaths per year), probably limited to North and South America. But let's take a closer look at these spiders.

Loxosceles species are small to medium-sized spiders with a body length of 6–20 mm. The forebody of the spider is light brown to brown, sometimes slightly reddish, while the hind body is cream white, yellowish, or light gray, usually without noticeable markings. The legs are brownish with regular hairs but without large bristles or spines. The spiders have only six eyes, which are arranged in three groups. *Loxosceles* species are easily recognizable by this eye arrangement. The color of the forebody is also distinctive, as a middle, usually U-shaped area is darker and runs into a line up to the base of the hind body. This vaguely resembles a violin in outline and gives the spiders another of their common names: violin spiders.

© Association for the Promotion of Spider Research 2024
W. Nentwig et al., *House Spiders - Worldwide*,
https://doi.org/10.1007/978-3-031-70448-2_23

However, many spiders appear to have this marking, so the eye arrangement is a much more reliable way to recognize *Loxosceles* (Fig. 23.1).

Loxosceles species are **nocturnal** and live hidden in cracks and crevices under stones, under rotting logs, and under peeling bark, where they line their retreats with

Fig. 23.1 The North American brown recluse *Loxosceles reclusa* (**top**) and the Mediterranean recluse *Loxosceles rufescens* (**bottom**), two very similar species. (*Photos* top Richard S Vetter, bottom Gordon Ackermann)

silk. Sometimes a few threads are spun out from the hole, which can then be expanded into an irregular tangle of threads. Prey is restrained by these spider threads, generating a vibration signal that encourages the spider to attack. Recluse spiders are quite successful at capturing anything that approaches them. Often mentioned are silverfish, cockroaches, termites, ants, flies, beetles, woodlice, and other spiders. As nocturnal hunters, recluse spiders have two strategies: they either lurk at night at the entrance area of their hideout or they roam freely in the surrounding area to find their prey.

In human settlements, recluse spiders can be found in woodpiles, under cardboard boxes, old tires, or in piles of rubbish. In buildings, they prefer to live in basements, in storage rooms, in the attic, in kitchen cabinets, behind skirting boards and picture frames, under bookshelves and dressers, in the sofa, basically anywhere where cracks and crevices in the wood or masonry offer them hiding places. Barns and all kinds of warehouses are also popular with recluse spiders. Behind wall cladding, under insulation, or under objects lying on the floor, they can occur **in high density**, usually completely unnoticed by humans.

The secretive life of *Loxosceles* can be clearly documented in numbers: in various houses in the United States, an average of 100 or more spiders were found in a short time through targeted searching by experts and with the help of glue traps. In one barn, the hunting success amounted to over 1000 recluse spiders within three nights. Despite this high *Loxosceles* density, <u>none</u> of the human inhabitants of the examined houses were ever bitten by these spiders.

Recluse spiders are not aggressive; bites are very rare, and some scientists speak of a real **bite inhibition** toward warm-blooded organisms. In fact, bites only occur when the spider fears being crushed. The classic bite scenario for *Loxosceles* therefore occurs when someone puts on a piece of clothing in which a spider is hiding or when they grab something where a spider is sitting. The spider's only defensive reaction is a bite, in about half of all cases, into the arms and legs of the victim. The unusual thing about a *Loxosceles* bite is that it is generally **painless**, whereas all

Fig. 23.2 The Brazilian brown recluse *Loxosceles intermedia* (**left**) and the Chilean recluse *Loxosceles laeta* (**right**). (*Photos* left Cesar Crash, right Gordon Ackermann)

other spider bites generally cause a sudden, sharp stinging pain, which is usually compared to a wasp sting. Therefore, there are no useful indications about the actual frequency of *Loxosceles* bites. Spiders are extremely economical with their **venom** and often resort to so-called "dry bites," i.e., no venom is injected. Experts, therefore, assume that most bites, apart from slight swelling with itching, cause no symptoms and heal on their own.

It is not clear why bites from these spiders are painless. Normally, the stinging pain of a spider bite is attributed to **histamine** in the venom, similar to wasp stings. However, *Loxosceles* species also have histamine in their venom, so more research is evidently needed here.

The venom of *Loxosceles* has another remarkable feature, because unlike all other spider venoms, its main active ingredient (instead of the usual neurotoxins) is an **enzyme**, sphingomyelinase D. This enzyme destroys cell membranes and causes cells to die. For this reason, when a recluse spider bites a human and injects venom, this can lead, after some time, to a narrowing of the blood vessels at the injection site, causing the tissue to become inflamed and die. Although this area usually heals on its own, the healing process is painful and takes weeks to months. In addition, the area of dying tissue can spread, such that surgical removal of the tissue may be advisable.

There has been intense discussion in the scientific literature for a long time about how often such severe **tissue destruction** occurs. The problem is that this damage only occurs after some time, and therefore a direct causal relationship to a spider bite, which moreover was painless, is difficult to establish. In medical practice, it has therefore become common to diagnose all possible complications of various bacterial skin infections that resemble such tissue destruction as *Loxosceles* spider bites, which unfortunately has led to a dramatic overestimation of their frequency. It is, therefore, important to note that in areas where *Loxosceles* species are not found or are uncommon, these spiders are the least likely cause of necrotic skin lesions, and other medical causes should be tested before blaming a recluse spider.

We now assume that *Loxosceles* bites resulting in serious and long-lasting tissue destruction are very rare events, which can be effectively treated today. However, in individual cases, a bite from *Loxosceles* can cause a severe systemic reaction, which can lead to the destruction of red blood cells (hemolysis), multiple organ failure, and death within a few days. It is unclear under what conditions a *Loxosceles* bite leads to such fatal consequences. Presumably, the problem of carrying out better research into such systemic cases is also related to the great rarity, as such deaths occur only a few times a year worldwide.

There are currently 143 *Loxosceles* species. They occur worldwide, but most species live in America (7% of all species in North America, 33% in Central America, 36% in South America), 20% occur in Africa, only 4% in Asia, and none in Australia. In general, you can encounter recluse spiders almost anywhere in the world. Strangely, bites are hardly known from Africa and Asia, and the specialist literature focuses on only four *Loxosceles* species. The most important species are probably the Brazilian brown recluse *Loxosceles intermedia* and the Chilean recluse

Box 23.1: That Glaring Point in Helsinki

When strange spiders were found in the Zoological Institute of the University of Helsinki in Finland in 1963, there was soon a murmur among the staff. Scientists had identified them as the Chilean recluse *Loxosceles laeta*, and external experts confirmed this diagnosis. So, a species of recluse spider that was previously only known from America and was also considered a "medically problematic species" was present. How on earth did such an animal come to Helsinki? And didn't they, in fact, need to inform and warn the general population and evacuate the building? Or at least thoroughly disinfect it? All these questions ultimately remained unanswered or lost their significance. For a long time, it was thought that the spiders had entered the building via box-by-box imports of apples from Argentina. However, this transport took place on refrigerated ships, which the spiders did not survive, as shown by later tests. An intensive search of neighboring buildings did not yield a single individual of *Loxosceles laeta*. Conversely, the attempt to empty the institute failed. An intensive hunt was followed days later by the discovery that the spider population had not decreased. The population obviously multiplied as expected but did not spread into the neighborhood. Above all, the animals proved to be nonaggressive. Even after 60 years, it was found that the animals were still present, had not spread, and no one had ever been bitten. So, for now, this isolated occurrence in Europe, that glaring point in Helsinki, remains on the map (Fig. 23.3).

Loxosceles laeta (Fig. 23.2, Box 23.1). It is known from Chile that *Loxosceles laeta* occurs in almost every household. Similarly, *Loxosceles intermedia* is widespread in Brazil (Fig. 23.3).

Fig. 23.3 Distribution areas of the Brazilian brown recluse *Loxosceles intermedia* (green), the Chilean recluse *Loxosceles laeta* (yellow-brown), and the brown recluse *Loxosceles reclusa* (blue)

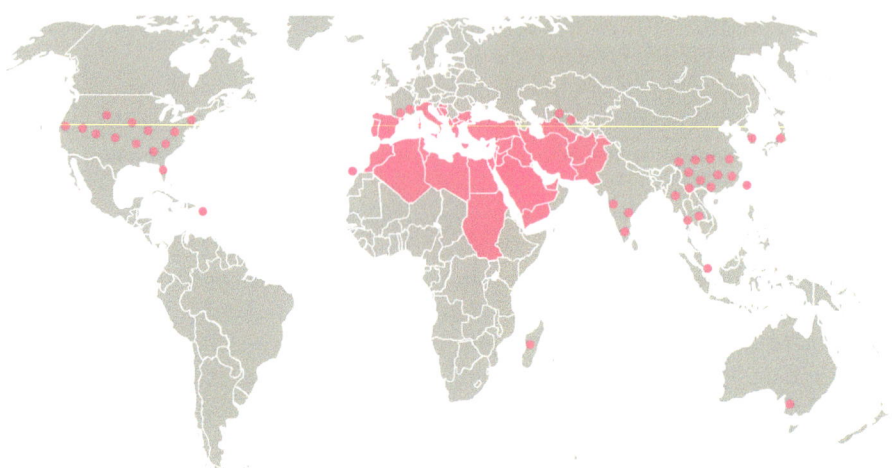

Fig. 23.4 Distribution area of the Mediterranean recluse *Loxosceles rufescens* (pink)

In North America, the brown recluse *Loxosceles reclusa* is considered less toxic and is limited to a southeastern area of the United States (Fig. 23.3). In Europe, on the other hand, only one species occurs in the Mediterranean area (the Mediterranean recluse *Loxosceles rufescens*) (Fig. 23.1), which was originally distributed from North Africa to Asia but has now been spread throughout the world (Fig. 23.4). Of the four medically important species, it is the one that causes the least problems in humans and has so far not caused any deaths.

We limit ourselves to the four *Loxosceles* species described above, which are regularly found in buildings and are medically significant. They are very similar in many aspects, so we treat them here together. Determining the species based on color alone is often difficult, as these species can vary in appearance. They can usually only be clearly distinguished by the microanatomical structure of their sexual organs, which we will not go into here.

Giant Crab Spiders (Sparassidae)

24

Giant crab spiders, also called huntsman spiders, with almost 1500 species, are a diverse family of predominantly large, free-living spiders. The similarity to crab spiders, which is reflected in their common name, is only superficial. The two families are not closely related. Their names just refer to similar movement patterns. Giant crab spiders have their legs directed sideways and usually move sideways, although they can also move forwards and backwards — and all at about the same speed. In this sense, they are similar to both crabs and crab spiders. In addition, sparassids can be recognized by their large, flat bodies and noticeably long legs. At the tips of the legs are dense pads of adhesive hairs (scopulae), which allow the spiders to move on vertical structures or under ceilings, even on smooth surfaces like glass.

Sparassids became famous as the family in which the **largest spiders in the world** are found. Now, "large" is a relative term when it comes to spiders. If it's about body length or body weight, this record surely belongs to a bulky tarantula, which at just under 10 cm body length weighs almost 200 g and has a leg span of 25 cm. However, when it comes to leg length or leg span, giant crab spiders are unbeatable. The largest sparassid reaches a body length of just under 5 cm and a leg span of about 30 cm. These are impressive dimensions but considerably less than some travel reports or horror films suggest. Moreover, this record-breaking giant crab spider lives in caves in Laos, and hardly anyone will ever see it there. The much more common giant crab spiders seen in houses, as presented here, are significantly smaller and reach leg spans of "only" about 12 cm.

Sparassids are found in the subtropics and tropics of all continents, with individual species also in the temperate zone. They live in a variety of habitats, often on rocks or trees, and hide during the day in crevices, cracks, and under bark. Sparassids are typically **nocturnal** and wait as **ambush predators** for passing prey.

Some species have adapted to life in human buildings. Since they are very successful at catching large cockroaches and other pests in our houses, they are actually welcome cohabitants. However, some people are quite "surprised" (to put it mildly) when, during the evening in a warm country, a "huge" spider suddenly appears on

W. Nentwig et al., *House Spiders - Worldwide*,
https://doi.org/10.1007/978-3-031-70448-2_24

the ceiling of their hotel room or holiday apartment. As we will see below, there is no reason to panic. Sparassids are harmless to humans and are shy and flee at any sign of approach.

24.1   Pantropical Giant Crab Spider *Heteropoda venatoria* (Giant Crab Spiders, Sparassidae)

This spider, which is nocturnal, "giant-sized," and can cause considerable "astonishment" with its jerky movements on house walls and ceilings, actually belongs in tropical Southeast Asia. Apparently, it has a preference for banana plants. This large spider can easily hide in their rolled-up leaves, under the dying parts, and especially between ripe bananas. And since **bananas** have always been taken by humans as a good, transportable food supply, they unwittingly spread these spiders with the fruit bunches to more and more areas.

With the advent of sailing ships and the colonial domination of the world, everything went even faster, so it is not surprising that *Heteropoda venatoria* now occurs on all continents (Fig. 24.2). Beyond the subtropical climate boundaries, it becomes too cold for them outdoors during some months, so they rely on the insides of buildings. Often these are air-conditioned warehouses, greenhouses, and buildings in botanical and zoological gardens. For example, in Germany, it has acquired the rather cumbersome common name "Warmhaus-Riesenkrabbenspinne" (warm house giant crab spider).

Heteropoda venatoria is a large spider, with females reaching a body length of 34 mm and males reaching 21 mm. The leg span reaches up to 120 mm in the largest animals, which is quite a lot for a spider but generally not very impressive in size. The body is flattened, as in typical crevice dwellers. The hind body is thick in females, especially when eggs are maturing, but in males, it is more slender.

The basic color of the body is light to medium brown. The female's forebody is often monochrome, while the male's forebody has a V-shaped delimitation that divides it into a light area bearing the eyes (Fig. 4.7) and two dark lateral areas. At the front edge of the eyes, both sexes have a whitish-yellow transverse stripe, which is a very good identification feature. The hind body shows a paired arrangement of light and dark spots, which are sometimes arranged in transverse lines. The legs show dark spots, especially in the area of the strong spines. Overall, this results in a spotted pattern for the spider, which camouflages it well, especially against tree bark, on branches, or on dry leaves (Fig. 24.1). Overall, however, it should be noted that the pantropical giant crab spider can vary greatly in color.

Heteropoda venatoria is a nocturnal ambush predator, meaning the spider leaves its daytime hiding place at dusk and waits for passing prey. The widely spread legs in their ambush position cover a wide area of the substrate, such that **vibrations** can be easily detected over several legs, and the direction from which they are coming can be perceived. The spider can feel where a prey animal is approaching from and can draw conclusions about its size and the type of prey. The prey spectrum is large, as almost all insects and other spiders are preyed upon. Due to its own body size, a

Fig. 24.1 The pantropical giant crab spider *Heteropoda venatoria*, female with eggsac in the chelicerae (**top**) and male (**bottom**). (*Photos* Guido Gabriel)

fully grown *Heteropoda venatoria* is capable of overpowering even large insects. In buildings, these include, for example, cockroaches, for which it seems to have a real preference, making it a welcome guest in many homes.

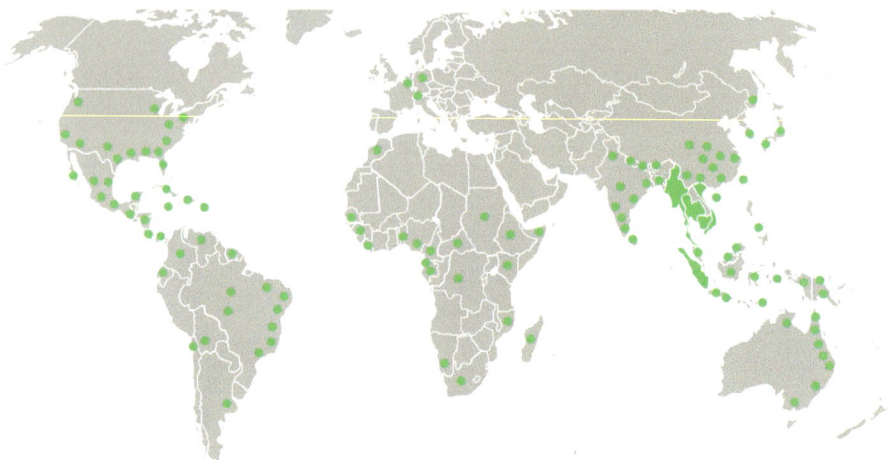

Fig. 24.2 Distribution area of the pantropical giant crab spider *Heteropoda venatoria* (green)

About four weeks after mating, the female produces a roundish, flat **eggsac**, which can have a diameter of up to 25 mm and is thus quite large. It contains about 200–300 eggs. During the four-week development period, the female holds the cocoon under her forebody with chelicerae and pedipalps (Fig. 24.1) and guards it until the young spiders hatch. This is the only time when the spider reacts aggressively to danger. After about a year, the young spiders are fully grown. The lifespan of the females can be up to two years.

The pantropical giant crab spider, like all species of this family, is considered **shy**. If it sees or feels the approach of something large that could potentially threaten it, such as a human, it flees. Because of their pronounced flight instinct, giant crab spiders are also difficult to catch. So, if you see a giant crab spider within your own four walls, you can rest assured that the spider will not come too close to you. The exceptions are females guarding an eggsac, who will try to defend themselves when threatened. Also, if you try to catch an animal with your hand or accidentally squeeze it, the spider may feel cornered and possibly bite.

If a **spider bite** should occur, *Heteropoda venatoria* is regarded as quite harmless. The bite causes immediate and intense pain, which, however, disappears after a few minutes. Sometimes, there may be skin redness and swelling at the bite site for a few hours, which may also bleed slightly. No further effects are known. Although giant crab spiders are large and powerful animals, they can therefore be classified as not dangerous. The biggest problem with *Heteropoda venatoria*, and related species, seems to be more psychological, namely that many people are afraid of this large and sometimes jerkily moving animal. We hope to have shown that there is no reason for this fear.

Occasionally, other Sparassidae species also occur in houses. In North America, you may find *Olios* species in the house; in North Africa and the Middle East, *Eusparassus walckenaeri* can occur; in southern Africa, *Palystes* species; and in Australia, species of the genera *Isopeda*, *Isopedella*, or *Neosparassus*.

Cobweb Spiders (Theridiidae)

Cobweb spiders, or Theridiidae, with just over 2500 species, are the fourth largest spider family. They are usually small to medium-sized species that can occur world-wide in all terrestrial habitats. The common name cobweb spider refers to their web type. Occasionally, they are also called tangle-web spiders, but this is a rather vague term (who knows exactly what this means) because there are many theridiids that build other types of webs or none at all.

Theridiids are web spiders and build a **capture web**, which consists of an irregular web cover, suspension threads, and other threads. The animals then sit upside down under the web cover and wait for prey. Many species have a retreat in which they can hide. These webs are long-term trapping devices that can be repaired and expanded. They also become soiled and occasionally need to be cleaned.

The webs of the cobweb spiders are remarkably effective insect traps, although they usually appear small and dirty. This is due, on the one hand, to the catching behavior of the spiders, and, on the other hand, to their sticky threads. Theridiids have modified spinnerets, from which they secrete droplets that are significantly more robust than those of orb-weavers. Thus, they can produce very effective **sticky threads**, which are usually located under the web and directed toward the ground. These trapping threads immediately tear off from the surface at the slightest touch of an insect. Passing insects bump into them, become stuck, and are then yanked upwards by the trapping thread toward the spider. This, of course, draws the spider's attention, which then demonstrates the special prey-catching strategy of cobweb spiders. It turns its rear end toward the prey and bombards it with large amounts of additional sticky threads. For this purpose, it has a striking, toothed bristle comb on the hind legs, and an alternative common name for this family is "comb-footed spiders." Using this comb and thread system, even prey animals that are significantly larger than the spider itself can be overpowered (Fig. 25.7).

Once the prey is immobilized, the spider looks for a thin spot in the insect's armor, which it can bite through with its small jaws. After an extensive meal, the spider appears significantly larger, as its hind body becomes massively swollen.

From the food supply in its gut, which can come from a single meal, it can live for several weeks or use this energy to produce many eggs.

Cobweb spiders are made for life in the web. With their long, thin legs, most species therefore appear quite clumsy outside their webs. The spiders have eight small eyes (Fig. 4.21) and are notable for their regular fine hairs on the legs. Theridiid legs lack spines, which distinguishes them from, for example, the sheet-weavers (Linyphiidae).

Cobweb spiders are not easy to identify, as, like in many spiders, color and patterning within a species can vary greatly and differences between closely related species can be small. Once again, we point out that for definitive species identification, the sexual organs would have to be used, which we do not do here.

Cobweb spiders are among the most common house spiders but lead a very secluded life. Therefore, the widow spiders, red house spiders, common house spiders, and false widows that we discuss here, among others, are usually hardly noticed in the house.

25.1 Widow Spiders of the Genus *Latrodectus* (Cobweb Spiders, Theridiidae)

Widow spiders have an awful reputation and are feared because they can be dangerous to humans. Moreover, they possess a considerable creepiness, as it is generally assumed that females eat their partners during mating (which is not always true but gives them their common name). Widow spiders (Theridiidae) are interesting creatures that are often beautifully colored. Wildlife photographers appreciate them as splendid photo subjects.

The best-known widow spiders are the **southern black widow** *Latrodectus mactans* (Fig. 25.1a, b) from North America; the **western black widow** *Latrodectus hesperus* (Fig. 25.1c), which is also North American; the **European black widow** *Latrodectus tredecimguttatus* (Fig. 25.2a), which is Eurasian; the **brown widow** *Latrodectus hesperus* (Fig. 25.2b), which is originally from Africa; and the **redback spider** *Latrodectus hasselti* (Fig. 25.2c), which is from Southeast Asia and Australia.

All widow spiders share the fact that they have spread far beyond their original distribution area through the transport of human goods, such that there are now widow spiders in almost all regions of the world, provided the climate suits them (Fig. 25.3). Thus, in Europe north of the Alps, *Latrodectus* species were regularly found associated with imported US sports cars or classic cars. Electronic or machine parts are also known for having such spiders traveling inside or outside their packaging. However, none of these newcomers have been able to establish themselves in northern Europe thus far.

Today, 35 *Latrodectus* species are known throughout the world. Although there have occasionally been reports of widow spiders in our homes, they are not actual house spiders. Rather, they are usually found outside our houses, on the outside of buildings, in neglected areas, garages, barns, sheds, outdoor toilets, and the like.

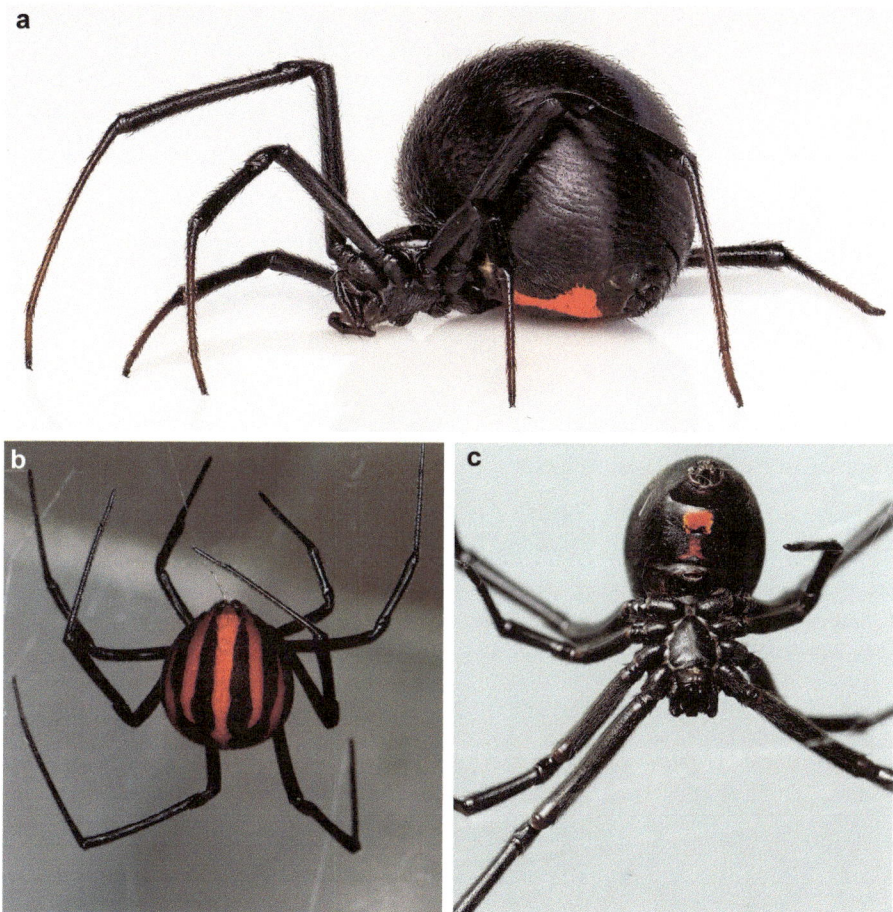

Fig. 25.1 (**a**) Female of the southern black widow *Latrodectus mactans*, (**b**) the same species in a Mexican variant with back markings, and (**c**) the western black widow *Latrodectus hesperus*. (*Photos* (**a**) Gilles Arbour/natureweb.com, (**b**) Gordon Ackermann, (**c**) Kyron Basu)

The spiders build their webs in dark corners and live on insects and other invertebrates that they can catch there.

Widow spiders, with their spherical hind bodies, have the form of classic cobweb spiders (Theridiidae). With a body length of 6–16 mm, they are medium-sized spiders. The legs are medium to long, typical for web-building spiders, and apart from numerous hairs, they do not have any noticeable spines or bristles. The basic color of most species is shiny black. The southern black widow *Latrodectus mactans* has two red spots on the underside that resemble an hourglass and usually a small red spot near the spinnerets. Additional red spots can appear on the back (Fig. 25.1a, b). The western black widow *Latrodectus hesperus* has a similar hourglass marking on the underside, usually red, sometimes yellow, or whitish (Fig. 25.1c). However, for

Fig. 25.2 (**a**) The European black widow *Latrodectus tredecimguttatus*, (**b**) the brown widow *Latrodectus geometricus*, and (**c**) the redback spider *Latrodectus hasselti*. (*Photos* (**a**) Eckhart Derschmidt, (**b**) Barbara Knoflach-Thaler, (**c**) Bastian Rast)

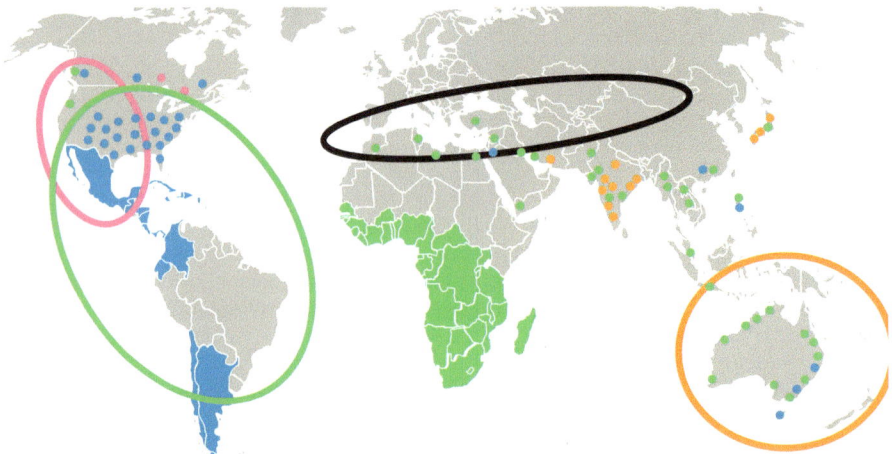

Fig. 25.3 The approximate distribution areas of the redback spider *Latrodectus hasselti* (yellow-brown), the western black widow *Latrodectus hesperus* (pink), the brown widow *Latrodectus geometricus* (green), the southern black widow *Latrodectus mactans* (blue), and the European black widow *Latrodectus tredecimguttatus* (black)

many individuals, an exact species identification is only possible by examining the sexual organs.

Also with an underlying black coloration, the redback spider *Latrodectus hasselti* exhibits an orange or red-colored stripe along its back (Fig. 25.2c), which in the

European black widow *Latrodectus tredecimguttatus* is broken down into 13 individual spots (Fig. 25.2a). Finally, the brown widow *Latrodectus geometricus* distinguishes itself by its underlying brown color, the red hourglass marking on the underside, and a very variable pattern of red spots on the back, often edged in white (Fig. 25.2b). Males are smaller in all species and usually differently colored. Males normally live for only one year, while females can live for 2–3 years.

In their natural habitat, these spiders mostly build a very messy ground-level web in dense vegetation or between large stones. In the central, densely woven part of the web, the spider has its retreat, into which it immediately hides when disturbed. Since this retreat is open at the bottom, the spider can escape downward in case of danger. From the retreat, *Latrodectus* spins threads in all directions, usually without a recognizable pattern. Interesting are those individual threads that reach the ground and are provided with adhesive drops. Insects become held by the adhesive drops, try to free themselves, and wriggle around wildly. The adhesive thread then tears off from the ground, and the glued insect is pulled up with the sticky thread, where it becomes even more entangled in further threads.

This whole action simultaneously triggers a strong **vibration signal**, which is transmitted to the spider via additional threads in the web. Based on the vibration pattern, the spider quickly decides whether it is a prey item or rather a threat and either strikes quickly or moves back deeper into its retreat. With this type of web, *Latrodectus* can catch smaller insects as well as medium-sized grasshoppers, cockroaches, beetles, bees, bumblebees, and ground-dwelling spiders. Such animals are then wrapped by the widow spider in silk, given a venomous bite, and sucked dry at leisure.

Judging by their reactions, widow spiders are **not aggressive**. In fact, they are somewhat timid. If you accidentally touch the web of a *Latrodectus* species, the spider in most cases realizes that this is not one of its typical prey items and retreats. Only under unusual circumstances does this lead to a bite. Perhaps your finger in the spider web is mistaken for a large insect, or you accidentally squeeze the spider because you pick up a leaf or another object under which it is sitting. Bites to humans are only caused by females, as the males are short-lived and too small.

The **venom** of *Latrodectus* species, unlike all previously studied spider venoms, has a special feature: it contains a group of neurotoxins characterized by their unusually large molecular weight. After a venomous bite, these attach themselves to the cell membrane of the nerve cells of their prey in such a way that they are perforated and irreversibly damaged.

Nevertheless, in about one-third of all bites, humans experience no symptoms or only mild pain that disappears within a few hours on its own. In such cases, the spider has not injected any venom, which happens more often than you might think. In all other cases, however, severe pain occurs, which can even prevent the bitten person from sleeping. The intensity of the pain typically increases during the first hour and radiates into the lower abdomen. Typical symptoms include sweating in 70% of cases and further systemic effects in 20–30% of cases (nausea or vomiting, increased temperature, high blood pressure). The pain lasts one to two days and other symptoms up to four days. **Symptomatic therapy** is recommended: painkillers for the pain, anti-nausea medication for nausea, and so on.

For very severe cases, an **antivenom** (also called an immunoserum) is available, i.e., antibodies against individual components of the spider venom that support the body's own immune system. Since such an antivenom often causes side effects (as it essentially contains foreign proteins that the body can fight), its use is limited to medically difficult cases. The availability of antivenoms has led to the fact that there have been (probably worldwide) no deaths from confirmed bites by *Latrodectus* species for over 50 years. These bites are still unpleasant, but with the help of modern medicine, they are no longer life-threatening.

Widow spiders are therefore much better than their reputation suggests. They are clearly not the deadliest spiders in the world, as a quick Internet search would suggest. And for the males of *Latrodectus* species, mating does not automatically end in death. According to various studies, there are species where the males always survive mating; in others, about eight to nine out of ten males survive. Not too bad a rate, right?

25.2   Red House Spider *Nesticodes rufipes* (Cobweb Spiders, Theridiidae)

Its web is usually located in a dark corner on the wall, under furniture, or on plant pots, and sometimes even in cabinets. It looks quite messy and randomly patched together. This web belongs to the red house spider *Nesticodes rufipes*, and, in addition to prey acquisition, it also serves as a retreat and a place of residence for the young. When disturbed, the spider simply drops but always remains secured by a safety thread connected to the web.

The red house spider is comparatively small, with a body length of only 3–6 mm, and females of this species are much larger than the males. The forebody and the hairy legs are shiny yellow-brown-red in color, such that the spider is quite noticeable. The hind body is spherically arched, pale without a clear pattern, but densely covered with short hairs. Sometimes individual guanine plates shimmer white, and occasionally a light central band is also visible. The degradation product guanine is usually excreted with the feces but can also be stored in crystal form under the thin cuticle, causing white spots to appear. The underlying color of the hind body is pink-light gray, occasionally slightly greenish, and sometimes dark (Fig. 25.4). The eight eyes are rather close together in two rows.

As a nocturnal species, *Nesticodes rufipes* feeds on all kinds of insects that get caught in its web, or that it also actively hunts at night on the walls. It particularly likes to eat houseflies, but ants and beetles are also welcome. Adult animals occur all year-round, a rather unusual occurrence among spiders. Therefore, mating and egg laying can occur at any time of the year. The round eggsacs are attached near the web. They are larger than the female's hind body and are noticeable for this reason alone.

This often motionless and therefore rather inconspicuous spider was first discovered by scientists based on specimens from Algeria and then found in more and more tropical countries around the world. This distribution posed puzzles, and in

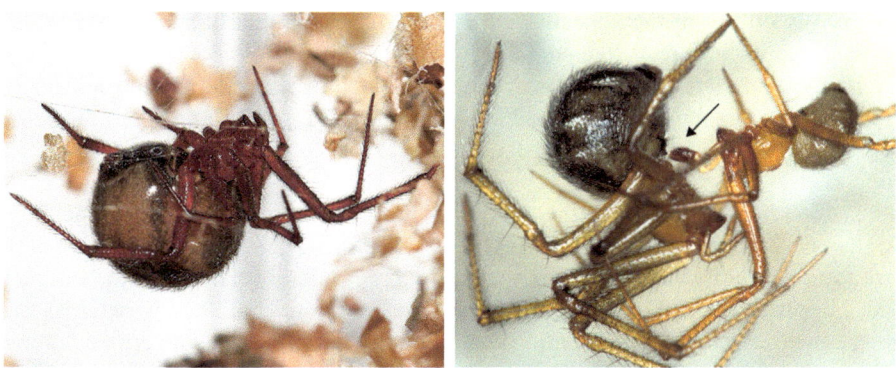

Fig. 25.4   The red house spider *Nesticodes rufipes*: **left**, a female; **right**, a pair immediately before copulation. The male inserts (from the right) his pedipalp into the sexual opening (epigyne) of the female (arrow). (*Photos* left Guido Gabriel, right Barbara Knoflach-Thaler)

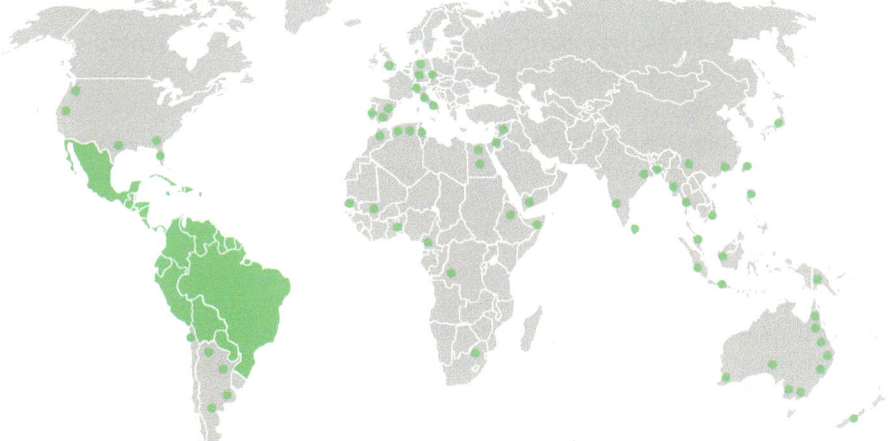

Fig. 25.5   The distribution area of the red house spider *Nesticodes rufipes* (green)

older literature, it was often described as "pantropical." However, a species cannot possibly have originated simultaneously across the entire tropics of the world (which is what pantropical means) but must have a continent of origin. Today we know that *Nesticodes rufipes* originally comes from Central and South America and, like so many other spiders, probably arrived with trade goods over the centuries by ships going from its home area to Europe and then the rest of the world (Fig. 25.5).

Apparently, it particularly likes pet stores, where it has been found several times in **insect farms** such as cricket farms. Laboratory ant farms are also very attractive to the red house spider. It significantly reduces the number of ants and can reproduce heavily in the ant nest. Such small-scale accumulations of insects apparently magically attract the spider. In Hawaii, this spider has colonized the small boxes that are laid out as glue traps against cockroaches. The space inside the box beneath the roof is safe from the adhesive, while cockroaches regularly approach. For the spider, it apparently feels like paradise in there.

Today, *Nesticodes rufipes* is predominantly found in tropical and subtropical climates on all continents of the world and colonizes all types of human buildings. In regions with cold winters, such as Central Europe, it is predominantly found in greenhouses.

The red house spider, like the widow spiders of the genus *Latrodectus*, belongs to the cobweb spider family (Theridiidae). Therefore, *Nesticodes* venom probably has some similarity to *Latrodectus* venom. It is occasionally claimed on the Internet that bites from the red house spider hurt for several hours but are otherwise harmless. However, there are no scientific studies on this. Even simple but verified reports of bites are completely missing. Moreover, *Nesticodes rufipes* is significantly smaller than the *Latrodectus* species, thus it could hardly bite us at all. Ultimately, it also shows a completely different behavior. It is extremely shy and prefers to retreat and hide. The red house spider, therefore, poses no threat to humans.

25.3   Common House Spider *Parasteatoda tepidariorum* (Cobweb Spiders, Theridiidae)

The common (sometimes called the American) house spider, *Parasteatoda tepidariorum*, originally comes from Asia and has now also spread to all parts of the world with the help of humans. As the Latin name suggests, a *tepidarium* was a warm room for the Romans, and the spider prefers to stay in warm environments. In some countries, it is thus called the greenhouse spider. It is found outdoors only in warmer areas of the world; otherwise, as in northern Central Europe, it is almost exclusively found inside heated buildings, such as **greenhouses**, nurseries, or flower shops.

This spider is quite inconspicuous in appearance. The forebody is uniformly light or medium brown. The hind body is, as is typical for the family, spherical, being higher than long. It has an underlying light brown color and shows irregular lighter and darker spots and lines. The legs are annulated light and dark brown. Males have a similar pattern but are more orange in color than the females (Fig. 25.6). With a body length of 3.5–7 mm in females and 3.5–6.5 mm in males, it is a relatively large, globular spider species.

The common house spider usually sits huddled in its large web, with its front legs angled over its body in such a way that it is barely visible. To the human eye, its web appears quite messy and chaotic, but it functions very efficiently as an **insect trap**. From the tangle of threads, individual capture threads lead downwards, which are equipped with highly effective adhesive droplets in their lower sections. Both insects running on the ground that bump into such a capture thread and flying insects that land in the web become hopelessly entangled.

To be on the safe side, the common house spider now uses the proven theridiid prey-capture technique, in which a trapped insect is bombarded with additional adhesive threads until it can no longer move. Only then is the prey killed with a venomous bite. The **venom** of the common house spider is very effective against

Fig. 25.6 The common house spider *Parasteatoda tepidariorum*: (**a**) female with eggsacs and hatching spiderlings, (**b**) female, and (**c**) female with male during courtship. (*Photos* (**a**, **c**) Gordon Ackermann, (**b**) Michael Hohner)

insects and spiders, so it is able to overpower animals that are significantly larger than itself, such as large beetles, bees, wasps, grasshoppers, crickets, or larger spiders (Fig. 25.7).

The common house spider does not use a retreat. However, if a leaf falls into its web, it gladly uses it as a hiding place. It usually places its web near the ground, such as on flower pots in winter gardens, but also in the transition joint between a wall and the ceiling, and preferably under window sills. If the web is placed high up in the room, the capture threads are not extended down to the ground but are drawn diagonally downwards toward the wall.

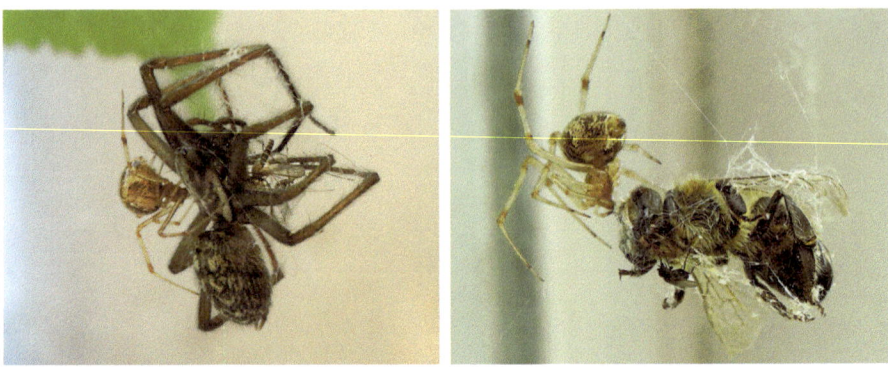

Fig. 25.7 The common house spider *Parasteatoda tepidariorum*: **left**, with the giant house spider (*Eratigena atrica*); **right**, with a honeybee as prey. (*Photos* Jutta Ansorg)

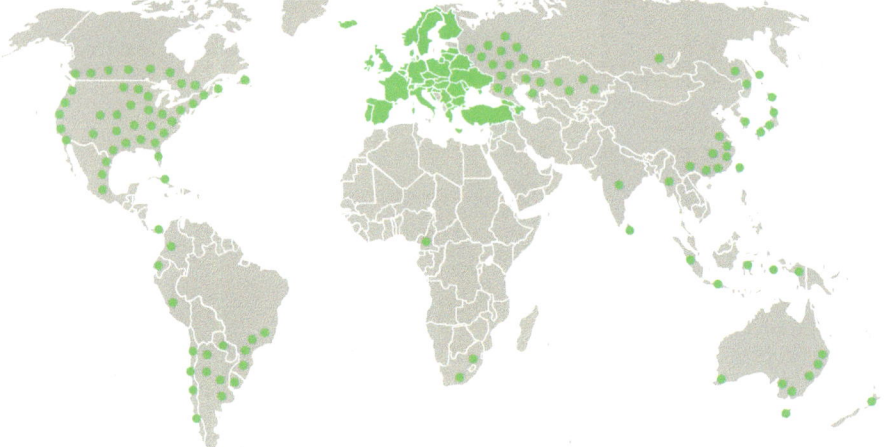

Fig. 25.8 The distribution area of the common house spider *Parasteatoda tepidariorum* (green)

In some areas, the common house spider is, as its name suggests, quite common. If there are enough prey animals available, many common house spiders can live in a small space, their webs then being placed relatively close to one another. Although there are habitats that the warmth-loving *Parasteatoda tepidariorum* does not like, it has been able to spread to all continents with human help (Fig. 25.8).

Before mating, females and males often live together in the web for several days. In the majority of cases, the male survives the mating and often then moves on to the next female. Mating itself is a brief affair and lasts only a few seconds. It is more astonishing that a female subsequently produces about three to eight **eggsacs**. These eggsacs have a gray-brown, paper-like shell, are about as large as the female's hind body, are pear-shaped, and are freely suspended in the web (Fig. 25.6). However, they are additionally protected from enemies with sticky threads. Each eggsac contains up to 250 eggs. Not only the eggsacs but also the newly hatched spiderlings are guarded by the mother until the young leave the web and seek their own habitats. After about 3 months and about seven molts, the spiderlings are fully grown.

The gray, pear-shaped eggsacs with their paper-like outer layer belonging to *Parasteatoda tepidariorum* (and other species of this genus) are clearly different from the eggsacs of *Steatoda* species, which can also be numerous in buildings. The latter are white, spherical, and fluffy on the outside.

25.4   False Widows of the Genus *Steatoda* (Cobweb Spiders, Theridiidae)

There are many species of the genus *Steatoda*, sometimes known as false widow spiders, which can be very difficult to distinguish from each other. To us humans, they appear overall brown and spherical, and since they usually hide in their retreat with their legs drawn in at the slightest vibration, they are often not noticed. Mostly, you only see their prey remains in and under the web because, like almost all cobweb spiders, false widows are extremely efficient insect catchers. In addition to flies, beetles, or woodlice, they even eat ants, which are avoided by many spider species. Even spiders larger than themselves are among their prey. We introduce three different species here, which have become associated with us humans worldwide.

In Central Europe, the **rabbit hutch spider** *Steatoda bipunctata* is particularly common. It has a dark brown, wrinkled forebody and reddish-brown, annulated legs. Its hind body is not as spherical as in other cobweb spiders but somewhat flattened. It has a greasy, shiny appearance as if it were hairless but is, in fact, covered with very fine, light hairs. The color of the hind body is medium to dark brown with a large, lighter area in the middle of the back, which actually looks more like a grease stain than a marking. The front edge of the hind body is bordered by a lighter line, and sometimes there is a lighter line visible in the middle of the hind body. Often, four dot-like depressions can be seen on the back, which represent attachment points for certain muscles (Fig. 25.9b). Males and females resemble one another in color, with the females being about 4.5–7 mm in body length, usually a bit larger than the males, which are about 4–5 mm in size.

Steatoda bipunctata is very common in houses (and in constructions like rabbit hutches) and prefers to weave its web in the corners of rooms and under window sills, such that it can hide well in a hole or a crack. It is a very undemanding species that also withstands the dry climate in our buildings well. But it also lives outdoors in suitable climates, on tree trunks or between rocks. The rabbit hutch spider is primarily found in Europe but also in North Africa, the Middle East, and Central Asia to China. It was introduced to North and South America. These findings in South America prove that the distribution area of the rabbit hutch spider is no longer limited to the northern hemisphere (Fig. 25.10).

It is interesting that males of the rabbit hutch spider, like all theridiids, can make sounds with their **stridulation apparatus**. This consists of a grooved area at the

Fig. 25.9 Female (**a**) of the triangulate cobweb spider *Steatoda triangulosa*, (**b**) of the rabbit hutch spider *Steatoda bipunctata*, and **c** of the large false widow *Steatoda grossa*. (*Photos* (**a**, **c**) Eckhard Derschmidt, (**b**) Jean-Philippe Taberlet)

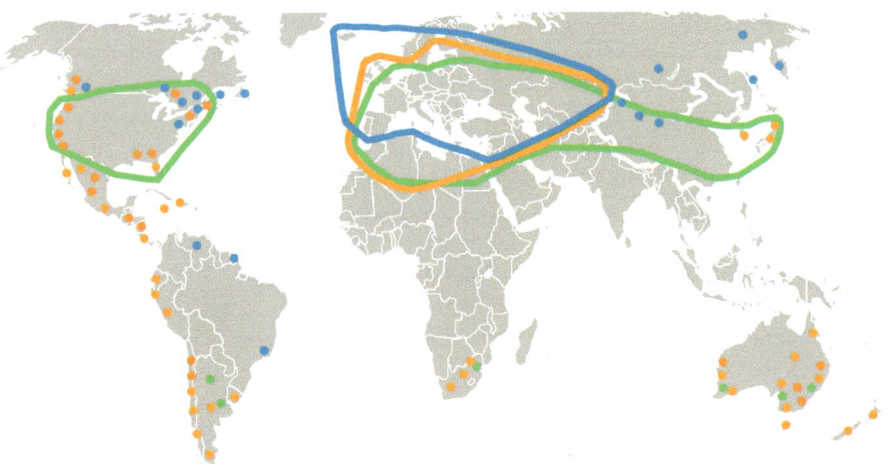

Fig. 25.10 The distribution areas of the triangulate cobweb spider *Steatoda triangulosa* (green), the large false widow *Steatoda grossa* (yellow-brown), and the rabbit hutch spider *Steatoda bipunctata* (blue)

rear edge of the forebody and strong spines located at the front edge of the hind body. These spines can be rasped over the groove-like area by moving the hind body up and down, creating a sound. The frequency of these sounds is around 1000 Hertz. This frequency is, in principle, audible to humans, but the volume is too low. The male uses these sounds during courtship to charm a female and impress other males.

A fertilized female produces several white eggsacs that are protected by enveloping adhesive threads. The **eggsacs** are spherical and fluffy and are attached to the edge of the web. Each eggsacs contains about 50–100 pink to purple-red eggs and is about the size of the female's hind body. By the way, you can also distinguish *Steatoda* from *Parasteatoda*, which is also very common in buildings, based on their eggsacs. The latter build pear-shaped, reddish-brown or gray-brown eggsacs that have a paper-like shell (Fig. 25.6).

In contrast to the rabbit hutch spider, the slightly larger **triangulate cobweb spider**, *Steatoda triangulosa*, not only has a more spherical hind body but is also clearly patterned. Some describe this pattern as a yellowish-white side and center band made of diamond spots of different sizes. Others see two dark brown zigzag bands on a cream-white background. The forebody is uniformly red to black-brown. The legs are yellowish-brown and darkly annulated (Fig. 25.9a). The coloration of the male is the same as that of the female. With a body length of 3.5–5 mm, it is usually smaller than the female, which can reach a body length of 3.5 to about 8.5 mm.

The triangulate cobweb spider is a cold-sensitive species that only occurs in the wild in warmer areas, such as southern Europe; otherwise, it only lives in buildings. It is native to Europe and has been able to expand its European distribution area in recent decades and has become cosmopolitan. It is more widespread in Asia than the rabbit hutch spider and has been introduced into Australia, Canada, the United States, Central and South America, as well as South Africa (Fig. 25.10).

The largest of the three species presented here is the **large false widow**, *Steatoda grossa*. Females can reach 6.5–10 mm in size, and males are usually not much smaller at 5–10 mm.

The coloration and patterning can be very variable but often resemble the triangulate cobweb spider, *Steatoda triangulosa*, which, however, is significantly smaller. In addition, the legs of the large false widow are always monochrome. The forebody is usually light brown like the legs, while the hind body can range from whitish-yellow to brown or even almost black (Fig. 25.9c). There are usually lighter rhombic spots on the hind body, which are, however, not as pronounced as in the triangulate cobweb spider and can also be completely absent.

Like the triangulate cobweb spider, the large false widow is cosmopolitan. The distribution areas of the two species differ little, but the large false widow, *Steatoda grossa*, has been found more frequently in Australia and in the west of South America than the triangulate cobweb spider. And unlike the other two species, it has also been found in New Zealand (Fig. 25.10).

Both the large false widow and the triangulate cobweb spider are capable of biting into human skin. The **venom** contains components that are also known from the true widow spiders of the genus *Latrodectus*, and occasionally warnings about the bite of false widows can be found on the Internet. In fact, no severe consequences from their bites are known. Nevertheless, a bite can certainly cause longer-lasting pain.

25.5   Black-Rumped Cobweb Spider *Theridion melanurum* (Cobweb Spiders, Theridiidae)

The black-rumped cobweb spider is a small, approximately 2.5–4.5 mm large theridiid, which can be found on house walls in almost all of Europe, parts of North Africa, the Middle East, and up into the Caucasus and Siberia. It was introduced to North America by humans (Fig. 25.12). It belongs to a group of several difficult-to-determine species, which can only be reliably identified by fine differences in their sexual organs. Furthermore, the species is very variable in color: In addition to relatively light animals with a characteristic zigzag-shaped central band on the hind body, there are also dark, almost black animals, and all conceivable transitional forms in between (Fig. 25.11).

In its natural environment, *Theridion melanurum* primarily inhabits crevices and cavities in rock walls. Like many other species from this habitat, it has found suitable substitute conditions in human structures. However, it is often overlooked there due to its small size and its finely spun **capture web**. The web contains capture threads mainly at the edges, upon which running and flying insects become stuck.

Then there is the special prey capture strategy of the cobweb spiders, which consists of bombarding a prey animal in the web with many sticky threads. In this way, even unusually large prey animals can be overwhelmed, and once they are immobilized, they receive the venomous bite. From such a huge prey animal, the spider's

Fig. 25.11 Female of the black-rumped cobweb spider *Theridion melanurum*: **left**, a dark colored animal; **right**, a light variant. (*Photos* Pierre Oger)

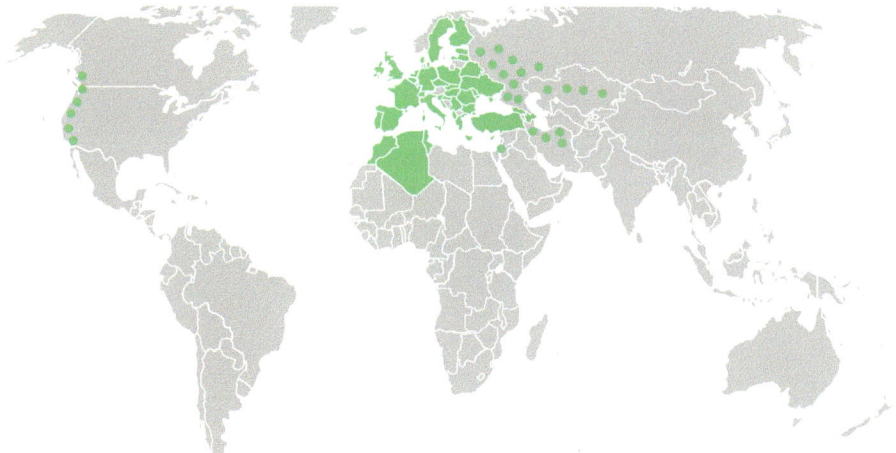

Fig. 25.12   The distribution area of the black-rumped cobweb spider *Theridion melanurum* (green)

hind body becomes thick and plump, and it can live on it for several weeks or use its nutritional energy to produce many eggs.

Spiders with an enlarged hind body are often females shortly before egg-laying. The spider then spins a fine, spherical, net-like, and brownish-marked **eggsac**, which contains about 100 gray-brown eggs. Once the eggsac is built, the female of *Theridion melanurum* spins a conspicuous protective web around the eggsac, which represents a wide, open-bottomed tube. In it, the female keeps watch next to its eggsac. It is thought that this protects the eggs from ants. But once again, this shows that we know very little about even common spiders in our immediate vicinity.

Hackled Orb-Weavers (Uloboridae)

With nearly 300 species, Uloboridae is a species-poor family of spiders, predominantly found in the subtropics and tropics, with only a few species reaching into the temperate zone. These predominantly small animals of 3–10 mm body length have, as a very special feature among spiders, completely reduced their **venom glands**. The reasons for this are unknown. It does mean, though, that they cannot envenomate captured prey, instead they must immediately wrap up a captured insect. Uloboridae have, therefore, become true **packaging artists** who wrap their prey so perfectly that escape is no longer possible. Apparently, the loss of venom glands has been more than adequately compensated in this way.

If you watch a hackled orb-weaver in its web, you will notice that it is also a master of camouflage. In many species, the forelegs are unusually long and strong. With these, they can easily adopt a stretched posture, in which the body, legs, and any dirt particles in the web merge into one unit, such that the spider itself is barely recognizable.

To the trained eye, it should be noted that the web threads look different from those of most other web-building spiders. Uloboridae are **cribellate** spiders, and their webs contain crimped threads (Chap. 1). Many uloborids are noticeable for their web type; several of them build **orb webs**. The orb webs of orb-weavers (Araneidae) are adorned with sticky threads and are medium to large in size, up to a meter in diameter in spiders like *Nephila*. By contrast, many hackled orb-weavers build small orb webs in the 5–20 cm range, which are typically horizontally oriented rather than vertical. The family also became famous because some species have reduced the orb web to a triangular sector of the web or even down to a single thread.

© Association for the Promotion of Spider Research 2024
W. Nentwig et al., *House Spiders - Worldwide*,
https://doi.org/10.1007/978-3-031-70448-2_26

26.1   Feather-Legged Hackled Orb-Weaver *Uloborus plumipes* (Hackled Orb-Weavers, Uloboridae)

At first glance, you might think you are seeing a dried-up, grayish-brown leaf that has accidentally landed on a spider web. If you try to grab it, it comes alive. It is the feather-legged hackled orb-weaver (*Uloborus plumipes*). In extremely strange positions, it can hang in its often horizontally built web but almost always holds the front pairs of legs stretched forward while it waits in its web for prey. This pose is typical for all members of the Uloboridae family, to which this quite peculiar but unmistakable spider belongs.

Females are about 3–6 mm in size, and males are 3–4 mm. The coloration of the spider is extremely variable. You can find yellow-greenish, red-brownish, or even grayish-black speckled specimens. Due to the dense hair, it appears almost fluffy (Fig. 26.1). Its body shape is striking. While the forebody is rather flat, the hind body is highly arched, triangular when viewed from the side, and often carries two whitish humps at the highest point. Particularly characteristic are the bushy, yellow-brown to black hairbrushes on the front legs of females. These gave the spider its scientific name because *plumipes* means "feathered at the feet."

Originally from the tropics and subtropics of Africa to the Near East, this rather small spider first spread to the Mediterranean area, then north of the Alps in **greenhouses** and garden centers, from which it has now also found its way into our homes. At the same time, it spread to Southeast Asia and Argentina (Fig. 26.2). Of course, the driving force behind this rapid spread was again global trade, this time mainly potted plants. Anyone who wants to observe this spider can easily find it in greenhouses and garden centers, as well as among the plants in the warm houses of botanical gardens.

There, this pretty, rather unusually shaped spider builds its mostly horizontal orb webs with cribellate capture threads, preferably between the large plant stems of potted plants, such as cacti, poinsettias, and many small shrubs. Webs can reach a

Fig. 26.1 Female of the feather-legged hackled orb-weaver *Uloborus plumipes* in motion (**left**) and in its camouflage position in the orb web (**right**). (*Photos* left Reinhard Bülte, right Barbara Knoflach-Thaler)

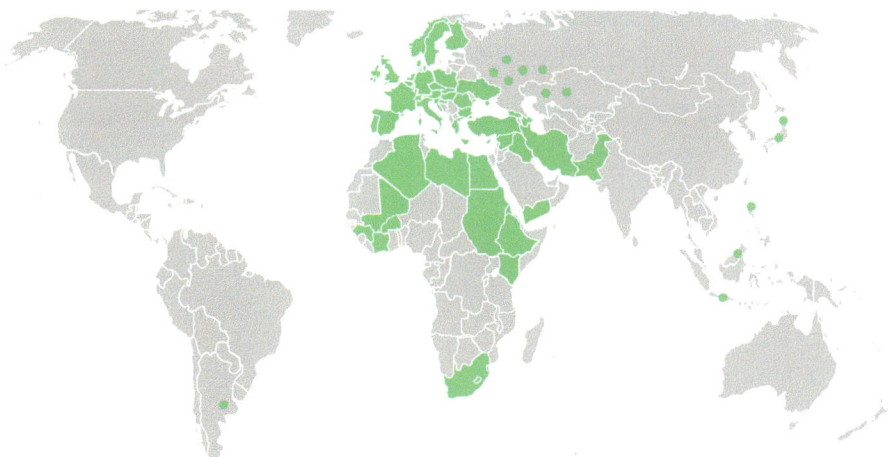

Fig. 26.2 Distribution area of the feather-legged hackled orb-weaver *Uloborus plumipes* (green)

considerable size of about 20 cm in diameter, such that the spider in the middle looks, in its elongated pose, rather like a shapeless stick or leaf.

Often one or more so-called **stabilimenta** (web bands) can be seen in the orb web. Their shape is very variable. It is usually an elongated band, often with a circular or spiral shape (particularly in juvenile webs), at least somewhat similar to the stabilimenta known from *Argiope* species (Chap. 7.2). When *Uloborus plumipes* adopts its camouflage position near such a silk band, it becomes even less conspicuous; thus, camouflage is probably a good explanation for the stabilimenta seen in *Uloborus* webs.

Since the feather-legged hackled orb-weaver does not possess venom glands and thus produces no venom to first paralyze and then kill its prey, it wraps the prey item very thoroughly until it is immobile. In this way, the spider is not bitten or stung by defensive insects out of carelessness. Prey is only bitten immediately before feeding. The **digestive enzymes**, which are then regurgitated into the prey and slowly dissolve it, ultimately fulfill the same function as venom and kill the prey. Often, one sees the spider afterward with a huge lump of prey and spider silk between the mouthparts.

Despite its small size, this spider consumes considerable numbers of small insects. The feather-legged hackled orb-weaver has proven to be very useful in controlling scale insects. It shields the breeding plants with its orb web and thus catches the pests before they can infest the plants. Whether *Uloborus plumipes* is the ideal organism for biological control, in other words, whether it is beneficial for controlling pests in greenhouses, may be doubted. Many customers do not want to buy plants with spider webs, and it does not seem sensible to accelerate the spread of this spider in this way.

In Europe, mating in *Uloborus plumipes* usually takes place in June or July, with egg-laying occurring a few weeks later. The elongated, yellowish-white **eggsac** is significantly larger than the female spider and is attached to the vegetation. A female can build several eggsacs. Adult animals can be found all year round. Since the small males were often overlooked, there was initially a suspicion that *Uloborus plumipes* might be a parthenogenetic species, meaning reproduction without males would be possible. However, upon closer investigation, it was found that this is not the case and both sexes are necessary for reproduction.

False Wolf Spiders (Zoropsidae) **27**

False wolf spiders are a small spider family, with only 181 known species. They resemble wolf spiders (hence the name) and hunt their prey freely without a web. However, unlike wolf spiders, they are **cribellate** spiders, meaning they produce a very fine, crimped spider silk known as cribellum wool (Chap. 1). However, this is no longer used to build webs for prey capture but is used only in the area of the retreat and to protect the eggs.

Zoropsidae species are medium to large spiders. They occur in many regions of the world and are, with one exception, limited to natural habitats. Generally, our knowledge about Zoropsidae is still very patchy. A remarkable exception is a species of the eponymous genus *Zoropsis*, which not only discovered a love for human buildings but has also experienced a strong expansion of its habitats. It will, therefore, be described in more detail below.

27.1 Mediterranean False Wolf Spider *Zoropsis spinimana* (False Wolf Spiders, Zoropsidae)

Encountering it can trigger very different emotions in humans, ranging from horror and fear to astonishment and joy. It can certainly happen that you go into the kitchen, bathroom, or bedroom in the evening, when it is already getting dark, turn on the light, and see a large, brown-black spider sitting on the wall. The animal does not seem surprised to be there and may give the impression that it feels it belongs here, like part of the furniture. It remains motionless in the middle of the light wall surface for the time being and then maybe takes a few slow steps to the side. We are facing *Zoropsis spinimana*, one of the largest European spider species.

Anyone in Europe who has paid even a little attention to spiders in the media in recent years will know this spider species because it has been a regular feature in the tabloid press for a while. The English name of this spider should be Mediterranean false wolf spider, which sounds a little boring. It is unclear who decided a few years ago to give this animal a catchier name, making reference to Nosferatu. Whoever it

W. Nentwig et al., *House Spiders - Worldwide*,
https://doi.org/10.1007/978-3-031-70448-2_27

was, strangely found that the pattern on the back of this spider resembles the face of the vampire from the eponymous 1922 horror film "Nosferatu – A Symphony of Horror" by Friedrich Wilhelm Murnau. This decision is difficult to comprehend, but now an (albeit extremely dubious) connection with vampires, bloodsuckers, and other horrors in the world was created. The harmless Mediterranean false wolf spider *Zoropsis spinimana* was suddenly supposed to be the dangerous "Nosferatu spider." But is there any connection between spiders and vampires? Or is this just a rather harmless and even useful spider being defamed out of ignorance? A character assassination of an otherwise honorable spider, as it were?

Where does the Mediterranean false wolf spider come from, and why did it suddenly appear north of the Alps in the 1990s? In essence, *Zoropsis spinimana* is also a victim of human-induced **climate change** because, originally, this spider was only limited to the western Mediterranean area. With increased warming, new habitats became available for this spider, and suddenly all of Central Europe and then more and more adjacent areas became climatically suitable. In 1994, it was sighted for the first time in Switzerland, in 1997 in Austria, and in 2005 in Germany. The march north seems, for now, to be unchecked. Cologne, Hamburg, and Berlin have long been reached in Germany, as have Belgium, Holland, and (in 2012) southern England. In the east, it has been sighted in the Czech Republic, Slovakia, and Poland. It is unclear how this spider could have spread so quickly, but it is likely that it was passively transported by vehicles. In 1992, *Zoropsis spinimana* was identified as an introduced species in California and thus new to North America (Fig. 27.2).

When talking about the habitat of the Mediterranean false wolf spider, another important point must be mentioned. Originally, it lived in the Mediterranean area on stones, rocks, and tree trunks, which also explains its ability to move so elegantly on vertical walls, as numerous adhesive hairs (scopulae) on the legs facilitate this **extreme climbing performance**. In the human cultural landscape, *Zoropsis spinimana* found ideal substitute habitats in our buildings, which are much more numerous than its previous environment. This also explains why the Mediterranean false wolf spider can effortlessly climb the exterior facade of a building up to, for example, the sixth floor. It then enters the interior of houses or apartments via balconies, windows, and doors, which is apparently not a pleasant thought for many people.

The Mediterranean false wolf spider reaches a body length of up to 19 mm (females) and 13 mm (males) and is, therefore, not particularly large. However, many people calculate things differently by adding the length of the legs. With a leg span of 6–8 cm, this results in sensational statements of 8 cm large spiders, which some people immediately assume without thinking to be the body length. The spider has robust legs, looks hairier than it actually is, and overall appears quite bulky (Fig. 27.1).

The animals are usually reddish or medium brown with a dark pattern on the back, which can be quite variable. On the back of the forebody, you can see a radial pattern, sometimes also described as a leaf-shaped mark. On the back of the hind body, the markings are usually interpreted as a series of three squares touching at their tips. These are dark brown to black and are clearly separated from the lighter sides of the hind body. Behind this, you can usually see some w-shaped lines. The

Fig. 27.1 Female (**left**) and male (**right**) of the Mediterranean false wolf spider *Zoropsis spini-mana*. (*Photos* Benjamin Eggs)

legs are usually darker at the tips and can be more or less clearly annulated or spotted. In their original habitat of rocks and tree bark, the spiders are very well camouflaged, but on white apartment walls, they stand out immediately.

Zoropsis spinimana is **nocturnal**, so a large part of its activity escapes our attention. During the day, it hides in a retreat, which is spun with a few slightly bluish shimmering silk threads. At night, it then sits somewhere in ambush, usually on a wall, and waits for passing prey. In this way, it can remain motionless and patient for hours. Laboratory tests have shown that the Mediterranean false wolf spider is polyphagous, meaning it can eat a very wide range of insects and other spiders. It is also capable of overpowering quite large prey. However, nothing is known about its diet under natural conditions.

In its lurking position, the Mediterranean false wolf spider appears completely motionless and apathetic, but it observes its surroundings very closely. The front side eyes and the rear middle eyes are slightly enlarged (Fig. 4.16), so that the spider can still see sufficiently in the dark. However, it receives the most important information primarily through its sensory hairs, which detect when a potential prey animal passes by. Once the prey item has been approached sufficiently, the spider grabs the insect with its two front legs. Only with particularly large or strongly wriggling animals do the two hind legs also come into play. The many adhesive hairs on the feet make it possible to hold a prey animal securely and firmly, as if the spider were glued to it. The erectable leg spines, which are particularly large on the front legs, also help to restrict the mobility of the prey, so there is no escape. In this lightning-fast sequence of movements, a targeted venomous bite is then made. Afterward, the spider releases its legs from the prey, holds it only with the mouthparts, and waits for the venom to take effect. So far, only a tenth of a second has passed. Now the spider takes its time and waits until later before proceeding to digest the prey.

The spider threads around its retreat and the silk with which it protects and guards its **eggsac** show that this is a cribellate spider, albeit one that no longer builds capture webs. Instead, it has become an ambush predator, behaving like an over-sized wolf spider, a spider family to which it is related.

Spiders mature in Central Europe in late summer; the main mating time is in autumn, and eggsacs with eggs are produced from winter to spring. A female can sequentially produce up to three eggsacs, which usually contain 20–50 eggs. The female sits on the eggsacs until the spiderlings hatch and defends it against threats as necessary. After the spiderlings hatch, it takes only eight months for them to develop into adult animals. The lifespan of females is at most one and a half years in total, while that of males is slightly shorter.

If you, as a spider friend, have discovered a Mediterranean false wolf spider in your apartment, you can live well together with it. You will hardly notice anything of the animal anyway, as it lives very hidden. For those who are kept awake at night by the thought of this, the simplest solution is to catch the spider and release it at some distance from your home. A sufficiently large glass or plastic container, which is slowly placed over the spider, is suitable for this. With a cardboard lid, you can close the container and remove it from the wall. If you now release the spider a few hundred meters further away, for example, in an area with industrial buildings, warehouses, or the like, you will surely never see the animal again.

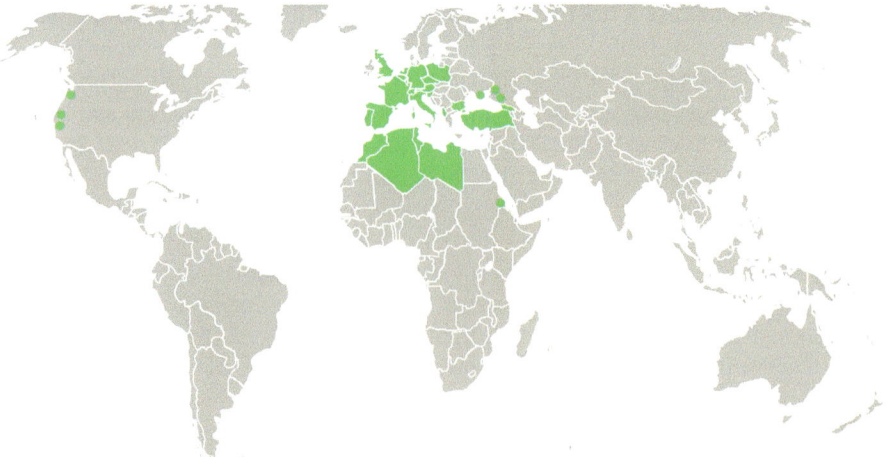

Fig. 27.2 Distribution area of the Mediterranean false wolf spider *Zoropsis spinimana* (green)

To catch *Zoropsis spinimana*, you need above all patience, as the spider is very shy. Often enough, the animal eludes our approach by fleeing. In such a hectic situation, you should not try to quickly catch the spider with your hands. In the first instance, you could injure the animal, but it is also more likely that *Zoropsis spinimana* will feel threatened by this action and do what it really dislikes doing: it bites. But don't worry, it's completely harmless.

Like almost all spiders, *Zoropsis* has a pair of venom glands that produce a small amount of **venom**. This venom is specifically designed for insects and acts quickly on them. Because the production of venom takes about two weeks and is laborious, spiders are not wasteful with their venom and only use as much as is absolutely necessary. So why use venom on humans, for whom they would never have a sufficient amount anyway and who they cannot eat? This is the only way to explain attempts by spider researchers who wanted to be bitten by *Zoropsis spinimana* to investigate the effects of this bite. In most cases, they failed because the spiders could not be persuaded to bite into the rather thick skin of an unattractive human.

If a bite does occur, more by chance and against the spider's will, the actual bite is hardly noticed or at most compared to the prick of a pair of tweezers. The pain that sets in afterward is comparable to a mosquito bite and rarely reaches the intensity of a wasp sting; thus, it is rather moderate. Sometimes a small redness or swelling follows, and that's about it. After a short time, the pain subsides and the skin changes to normal again. In very rare cases, there may be prolonged redness, swelling, and blistering, which, however, disappear after a few days. Reports to the contrary, which detail much worse consequences, are generally not credible.

And another point of reassurance is appropriate: *Zoropsis spinimana* is an ambush predator and does not run around on our bed when we are asleep. In addition to this, the spider is aware of our breathing movements through its vibration sense and therefore avoids sleeping people.

Literature

Overview Works

Beccaloni J (2009) Arachnids. Natural History Museum, London
Bristowe WS (1958) The world of spiders. Collins, London
Cowles J (2018) Amazing arachnids. Princeton University Press, Princeton
Dalton S (2008) Spiders: the ultimate predators. Bloomsbury, London
Foelix RF (2011) Biology of spiders, 3rd edn. Oxford University Press, New York
Herberstein ME (2011) Spider behaviour: flexibility and versatility. Cambridge University Press, Cambridge
Nentwig W (1987) Ecophysiology of spiders. Springer, Berlin
Nentwig W (2013) Spider ecophysiology. Springer, Heidelberg
Nentwig W, Ansorg J, Bolzern A, Frick H, Ganske A-S, Hänggi A, Kropf C, Stäubli A (2022) All you need to know about spiders. Springer, Cham
Platnick NI (2020) Spiders of the world. A natural history. Princeton University Press, Princeton

Internet Resources

American Arachnological Society. https://www.americanarachnology.org
Arachnologische Gesellschaft (Germany, Austria, Switzerland). https://arages.de
British Arachnological Society. https://britishspiders.org.uk
Dippenaar-Schoeman AS, Haddad CR, Foord SH, Lotz LN (2020–2023) South African National Survey of arachnida photo identification guide, Irene. https://zenodo.org/search?q=%22South%20African%20National%20Survey%20of%20Arachnida%20Photo%20Identification%20Guide%22&l=list&p=1&s=10&sort=-publication_date
Forum europäischer Spinnentiere. https://forum.arages.de
Global Biodiversity Information Facility (GBIF). https://www.gbif.org
International Society of Arachnology. https://arachnology.org
Metzner H (2023) Jumping spiders (Arachnida: Araneae: Salticidae) of the world. Online at https://www.jumping-spiders.com
Nentwig W, Blick T, Bosmans R, Gloor D, Hänggi A, Kropf C (2024) Araneae. Spiders of Europe. https://araneae.nmbe.ch/?lang=en
World Spider Catalog (2023) Version 23.5. Natural History Museum Bern, http://wsc.nmbe.ch. DOI 10.24436/2

© Association for the Promotion of Spider Research 2024
W. Nentwig et al., *House Spiders - Worldwide*,
https://doi.org/10.1007/978-3-031-70448-2

Identification Books

Bee L, Oxford G, Smith H (2017) Britain's spiders. A field guide. Princeton University Press, Princeton

Bellmann H (2016) Der Kosmos Spinnenführer. Kosmos, Stuttgart

Bradley RA, Buchanan S (2013) Common spiders of North America. University of California Press, Oakland

Framenau VW, Baehr BC, Zborowski P (2014) A guide to the spiders of Australia. New Holland Publishers, London

Jocqué R, Dippenaar-Schoeman AS (2006) Spider families of the world. Royal Museum for Central Africa, Tervuren

Le Peru B (2011) The spiders of Europe, a synthesis of data: volume 1 Atypidae to Theridiidae. Mem Soc Linn Lyon 2:1–522

Roberts MJ (1995) Collins field guide — spiders of Britain & northern Europe. Harper Collins Publishers, New York

Rose S (2022) Spiders of North America. Princeton University Press, Princeton

Ubick D, Paquin P, Cushing PE, Roth V (2005) Spiders of North America: an identification manual. American Arachnological Society, Keene

Vink CJ (2015) Spiders of New Zealand. New Holland Publishers, Auckland

Wiehle H (1931) Spinnentiere oder Arachnoidea. 27. Familie. Araneidae. Die Tierwelt Deutschl 23:47–136

Wiehle H (1937a) Spinnentiere oder Arachnoidea. 26. Familie. Theridiidae oder Haubennetzspinnen (Kugelspinnen). Die Tierwelt Deutschl 33:119–222

Wiehle H (1953) Spinnentiere oder Arachnoidea (Araneae) IX: Orthognatha-Cribellatae-Haplogynae-Entelegynae (Pholcidae, Zodariidae, Oxyopidae, Mimetidae, Nesticidae). Die Tierwelt Deutschl 42:1–150

Individual Works (Chapter by Chapter)

Chapter 5: Agelenidae

Ayoub NA, Riechert SE, Small RL (2005) Speciation history of the North American funnelweb spiders, Agelenopsis (Araneae: Agelenidae): phylogenetic inferences at the population-species interface. Mol Phylogen Evol 36:42–57

Binford GJ (2002) An analysis of geographic and intersexual chemical variation in venoms of the spider Tegenaria agrestis (Agelenidae). Toxicon 39:955–968

Bolzern A, Burckhardt D, Hänggi A (2013) Phylogeny and taxonomy of European funnel-web spiders of the Tegenaria – Malthonica complex (Araneae: Agelenidae) based upon morphological and molecular data. Zool J Linnean Soc 168:723–848

Guarisco H (2014) The funnelweb spider genus Agelenopsis (Araneae: Agelenidae) in Kansas. Trans Kansas Acad Sci 117:79–87

Nentwig W, Gnädinger M, Fuchs J, Ceschi A (2013a) A two year study of verified spider bites in Switzerland and a review of the European spider bite literature. Toxicon 73:104–110

Oxford G (2023) The identification of members of the Eratigena atrica group of large house spiders — E. atrica, E. duellica and E. saeva (Agelenidae). Newsl Br Arachnol Soc 156/SRS News 105:1–4

Roth VD (1968) The spider genus Tegenaria in the Western Hemisphere (Agelenidae). Am Mus Novit 2323:1–33

Whitman-Zai J, Francis M, Geick M, Cushing PE (2015) Revision and morphological phylogenetic analysis of the funnel web spider genus Agelenopsis (Araneae: Agelenidae). J Arachnol 43:1–25

Chapter 6: Amaurobiidae

Bott RA, Baumgartner W, Bräunig P, Menzel F, Joel A-C (2017) Adhesion enhancement of cribellate capture threads by epicuticular waxes of the insect prey sheds new light on spider web evolution. Proc R Soc B 284:20170363

Kim KW, Horel A (1998) Matriphagy in the spider *Amaurobius ferox* (Araneidae, Amaurobiidae): an example of mother-offspring interactions. Ethology 104:1021–1037

Kim KW, Roland C, Horel A (2000) Functional value of matriphagy in the spider *Amaurobius ferox*. Ethology 106:729–742

Kovács G, Szinetár C (2020) Adatok a nagy eretnekpók (*Amaurobius ferox* [Walckenaer, 1830]) biológiájához (Araneae: Amaurobiidae) [Data on the biology of *Amaurobius ferox* (Walckenaer, 1830) (Araneae: Amaurobiidae)]. Biológia Savaria Természettudományi és Sporttudományi Közlemények 18:75–92

Leech RE (1972a) A revision of the Nearctic Amaurobiidae (Arachnida: Araneida). Mem Entomol Soc Can 84:1–182

Tahiri A, Horel A, Krafft B (1989) Preliminary study on the interactions between mother-young and young-young in two species of *Amaurobius* (Araneae Amaurobiidae). Rev Arachnol 8:115–128

Chapter 7: Araneidae

Bjørn PP (1997) A taxonomic revision of the African part of the orb-weaving genus *Argiope* (Araneae: Araneidae). Entomol Scand 28:199–239

Blanke R, Merklinger F (1983) Die Variabilität von Zeichnungsmuster und Helligkeit des Abdomens bei *Araneus diadematus* Clerck und *Araneus marmoreus* Clerck (Arachnida: Araneae). J Zool Syst Evol Res 20:63–75

Davies VT (1988) An illustrated guide to the genera of orb-weaving spiders in Australia. Mem QLD Mus 25:273–332

Dondale CD, Redner JH, Paquin P, Levi HW (2003) The insects and arachnids of Canada. Part 23. The orb-weaving spiders of Canada and Alaska (Araneae: Uloboridae, Tetragnathidae, Araneidae, Theridiosomatidae). NRC Research Press, Ottawa, 371 pp

Harvey MS, Austin AD, Adams M (2007) The systematics and biology of the spider genus *Nephila* (Araneae: Nephilidae) in the Australasian region. Invertebr Syst 21:407–451

Heiling AM, Herberstein ME (1999) Asymmetry in spider orb-webs: a result of experience? Anim Cogn 2:171–177

Heiling AM, Herberstein ME (2000) Interpretations of orb-web variability: a review of past and current ideas. Ekológia (Bratislava) 19 (Suppl. 3):97–106

Kuntner M (2007) A monograph of *Nephilengys*, the pantropical 'hermit spiders' (Araneae, Nephilidae, Nephilinae). Syst Entomol 32:95–135

Kuntner M, Arnedo MA, Trontelj P, Lokovsek T, Agnarsson I (2013) A molecular phylogeny of nephilid spiders: evolutionary history of a model lineage. Mol Phylogen Evol 69:961–979

Kuntner M, Hamilton CA, Cheng RC, Gregorič M, Lupse N, Lokovsek T, Lemmon EM, Lemmon AR, Agnarsson I, Coddington JA, Bond JE (2019) Golden orbweavers ignore biological rules: phylogenomic and comparative analyses unravel a complex evolution of sexual size dimorphism. Syst Biol 68:555–572

Levi HW (1968) The spider genera *Gea* and *Argiope* in America (Araneae: Araneidae). Bull Mus Comp Zool 136:319–352

Levi HW (1971) The *diadematus* group of the orb-weaver genus *Araneus* north of Mexico (Araneae: Araneidae). Bull Mus Comp Zool 141:131–179

Levi HW (1974a) The orb-weaver genera *Araniella* and *Nuctenea* (Araneae: Araneidae). Bull Mus Comp Zool 146:291–316

Levi HW (1974b) The orb-weaver genus *Zygiella* (Araneae: Araneidae). Bull Mus Comp Zool 146:267–290

Levi HW (1983) The orb-weaver genera *Argiope*, *Gea*, and *Neogea* from the western Pacific region (Araneae: Araneidae, Argiopinae). Bull Mus Comp Zool 150:247–338

Nentwig W (1985a) Prey analysis of four species of tropical orb-weaving spiders (Araneae: Araneidae) and a comparison with araneids of the temperate zone. Oecologia 66:580–594

Nentwig W, Spiegel H (1986) The partial web renewal behaviour of *Nephila clavipes* (Araneae, Araneidae). Zool Anz 216:351–356

Robinson MH, Robinson B (1973) Ecology and behavior of the giant wood spider *Nephila maculata* (Fabricius) in New Guinea. Smithsonian. Contr Zool 149:1–76. (Refers to *Nephila pilipes*)

Šestáková A, Marusik YM, Omelko MM (2014) A revision of the Holarctic genus *Larinioides* Caporiacco, 1934 (Araneae: Araneidae). Zootaxa 3894:61–82

Zschokke S (2011) Spiral and web asymmetry in the orb webs of *Araneus diadematus* (Araneae, Araneidae). J Arachnol 39:358–362

Chapter 8: Cheiracanthiidae

Bryant EB (1951) Redescription of *Cheiracanthium mildei* L. Koch, a recent spider immigrant from Europe. Psyche 58:120–123

Corrigan JE, Bennett RG (1986) Predation by *Cheiracanthium mildei* (Araneae, Clubionidae) on larval *Phyllonorycter blancardella* (Lepidoptera, Gracillariidae) in a greenhouse. J Arachnol 15:132–134

Edwards RJ (1958) The spider subfamily Clubionidae of the United States, Canada and Alaska (Araneae: Clubionidae). Bull Mus Comp Zool 118:365–436

Foradori MJ, Smith SC, Smith E, Wells RE (2005) Survey for potentially necrotizing spider venoms, with special emphasis on *Cheiracanthium mildei*. Comp Biochem Physiol Part C 141:32–39

Taylor RM, Bradley RA (2009) Plant nectar increases survival, molting, and foraging in two foliage wandering spiders. J Arachnol 37:232–237

Vetter RS, Isbister GK, Bush SP, Boutin LJ (2006) Verified bites by yellow sac spiders (genus *Cheiracanthium*) in the United States and in Australia: where is the necrosis? Am J Trop Med Hyg 74:1043–1048

Chapter 9: Desidae

Costa FG (1993) Cohabitation and copulation in *Ixeuticus martius* (Araneae, Amaurobiidae). J Arachnol 21:258–260

Haddad CR, Foord SH (2021) Future climate may limit the spread of the Australian house spider *Badumna longinqua* (Araneae: Desidae) in South Africa. J Arachnol 49:332–339

Kielhorn K-H, Rödel I (2011) *Badumna longinqua* introduced to Europe (Araneae: Desidae). Arachnol Mitt 42:1–4

Leech RE (1972b) A revision of the Nearctic Amaurobiidae (Arachnida: Araneida). Mem Entomol Soc Can 84:1–182

Marples RR (1959) The dictynid spiders of New Zealand. Trans Proc R Soc NZ 87:333–361

Nentwig W (2015) Introduction, establishment rate, pathways and impact of spiders alien to Europe. Biol Invas 17:2757–2778

Simó M, Laborda A, Jorge C, Guerrero JC, Alves Dias M, Castro M (2011) Introduction, distribution and habitats of the invasive spider *Badumna longinqua* (L. Koch, 1867) (Araneae: Desidae) in Uruguay, with notes on its world dispersion. J Nat Hist 45:1637–1648

Chapter 10: Dictynidae

Billaudelle H (1957) Zur Biologie de Mauerspinne *Dictyna civica* (H. Luc.) (Dictynidae, Araneida). Z Angew Entomol 41:475–512

Chamberlin RV, Gertsch WJ (1958) The spider family Dictynidae in America north of Mexico. Bull Am Mus Nat Hist 116:1–152

Eberhard WG (2020) The web of *Dictyna bellans* (Araneae: Dictynidae). J Arachnol 48:272–277

Chapter 11: Dysderidae

Deeleman-Reinhold CL, Deeleman PR (1988) Revision of the Dysderinae (Araneae, Dysderidae), excluding the western Mediterranean species. Tijdschr Entomol 131:141–269

Pollard SD (1986) Prey capture in *Dysdera crocata* (Araneae: Dysderidae), a long fanged spider. NZ J Zool 13:139–150

Pollard SD, Jackson RR, Van Olphen A (1995) Does *Dysdera crocata* (Araneae Dysderidae) prefer woodlice as prey? Ethol Ecol Evol 7:271–275

Řezáč M, Král J, Pekár S (2008) The spider genus *Dysdera* (Araneae, Dysderidae) in Central Europe: revision and natural history. J Arachnol 35:432–462

Vetter RS, Isbister GK (2006) Verified bites by the woodlouse spider, *Dysdera crocata*. Toxicon 47:826–829

Chapter 12: Filistatidae

Barrantes G, Ramírez MJ (2013) Courtship, egg sac construction, and maternal care in *Kukulcania hibernalis*, with information on the courtship of *Misionella mendensis* (Araneae, Filistatidae). Arachnology 16:72–80

Curtis JT, Carrell JE (1999) Social behaviour by captive juvenile *Kukulcania hibernalis* (Araneae: Filistatidae). Bull Br Arachnol Soc 11:241–246

Magalhaes ILF, Ramírez MJ (2019) The crevice weaver spider genus *Kukulcania* (Araneae: Filistatidae). Bull Am Mus Nat Hist 426:1–151

Magalhaes ILF, Aharon S, Ganem Z, Gavish-Regev E (2022) A new semi-cryptic *Filistata* from caves in the Levant with comments on the limits of *Filistata insidiatrix* (Forsskål, 1775) (Arachnida: Araneae: Filistatidae). Eur J Taxon 831:149–174

Marusik YM, Zonstein SL (2014) A synopsis of Middle East *Filistata* (Aranei: Filistatidae), with description of new species from Azerbaijan. Arthropoda Selecta 23:199–205

Chapter 13: Gnaphosidae

Bolzern A, Hänggi A (2006) *Drassodes lapidosus* and *Drassodes cupreus* (Araneae: Gnaphosidae) — an endless story. Arachnol Mitt 31:16–22

Christian S (2015) *Micaria subopaca* Westring, 1861, *Scotophaeus blackwalli* (Thorell, 1871) and *S. scutulatus* (L. Koch, 1866): three species of gnaphosids new to Luxembourg (Arachnida, Gnaphosidae). Bull Soc Natural Luxembg 117:87–90

Grimm U (1985) Die Gnaphosidae Mitteleuropas (Arachnida, Araneae). Abh Naturwiss Ver Hamburg (NF) 26:1–318

Guarisco H (2007) Checklist of Kansas ground spiders. The Kansas School Naturalist, Emporia

Kaston BJ (1965) Some little known aspects of spider behavior. Am Midland Nat 73:336–356

Ortega-Escobar J (2017) Polarized-light vision in spiders. Trends Entomol 13:25–34

Platnick NI, Murphy JA (1984) A revision of the spider genera *Trachyzelotes* and *Urozelotes* (Araneae, Gnaphosidae). Am Mus Novit 2792:1–30
Platnick NI, Shadab MU (1977) A revision of the spider genera *Herpyllus* and *Scotophaeus* (Araneae, Gnaphosidae) in North America. Bull Am Mus Nat Hist 159:1–44

Chapter 14: Linyphiidae

van Helsdingen PJ (1965) Sexual behaviour of *Lepthyphantes leprosus* (Ohlert) (Araneida, Linyphiidae), with notes on the function of the genital organs. Zool Meded 41:15–42
Wiehle H (1956) Spinnentiere oder Arachnoidea (Araneae). 28. Familie Linyphiidae — Baldachinspinnen. Die Tierwelt Deutschl 44:1–337
Wiehle H (1960) Spinnentiere oder Arachnoidea (Araneae). XI. Micryphantidae — Zwergspinnen. Die Tierwelt Deutschl 47:1–620
Zorsch HM (1937) The spider genus *Lepthyphantes* in the United States. Am Midl Nat 18:856–898

Chapter 15: Mimetidae

Czajka M (1963) Unknown facts of the biology of the spider *Ero furcata* (Villers) (Mimetidae, Araneae). Pol J Entomol 33:229–231
Harms D, Harvey MS (2009) Australian pirates: systematics and phylogeny of the Australasian pirate spiders (Araneae: Mimetidae), with a description of the Western Australian fauna. Invertebr Syst 23:231–280
Thaler K, van Harten A, Knoflach B (2004) Pirate spiders of the genus *Ero* C.L. Koch from southern Europe, Yemen, and Ivory Coast, with two new species (Arachnida, Araneae, Mimetidae). Denisia 13:359–368

Chapter 16: Oecobiidae

Glatz L (1967) Zur Biologie und Morphologie von *Oecobius annulipes* Lucas (Araneae, Oecobiidae). Z Morph Tiere 61:185–214. (Misidentification, refers to *Oecobius navus*)
Santos AJ, Gonzaga MO (2003) On the spider genus *Oecobius* Lucas, 1846 in South America (Araneae, Oecobiidae). J Nat Hist 37:239–252
Wunderlich J (1995) On the taxonomy and biogeography of the species of the genus *Oecobius* Lucas 1846, with new descriptions from the Mediterranean and from the Arabian peninsula (Arachnida: Araneae: Oecobiidae). Beitr Araneol 4:585–608

Chapter 17: Oonopidae

de Dalmas R (1916) Revision of the genus *Orchestina* E.S., followed by the description of new species of the genus *Oonops* and a study on the Dictynidae of the genus *Scotolathys*. Ann Soc Entomol Fr 85:203–258
Korenko S, Smerda J, Pekar S (2009) Life-history of the parthenogenetic oonopid spider, *Triaeris stenaspis* (Araneae: Oonopidae). Eur J Entomol 106:217–223
Kupryjanowicz J, Staręga W (1994) *Oonops domesticus* Dalmas, 1916 – a new spider species for the fauna of Poland (Araneae: Oonopidae). Bull Pol Acad Sci Biol Sci 42:83–86
Platnick NI, Dupérré N (2009) The goblin spider genus *Heteroonops* (Araneae, Oonopidae), with notes on *Oonops*. Am Mus Novit 3672:1–72

Chap. 18 Pholcidae

Aharon S, Huber BA, Gavish-Regev E (2017) Daddy-long-leg giants: revision of the spider genus *Artema* Walckenaer, 1837 (Araneae, Pholcidae). Eur J Taxon 376:1–57

Huber BA (1994) Genital morphology, copulatory mechanism and reproductive biology in *Psilochorus simoni* (Berland, 1911) (Pholcidae; Araneae). Neth J Zool 44:85–99

Huber BA (2002) Functional morphology of the genitalia in the spider *Spermophora senoculata* (Pholcidae, Araneae). Zool Anz 241:105–116

Huber BA (2005) Revision of the genus *Spermophora* Hentz in Southeast Asia and on the Pacific islands, with descriptions of three new genera (Araneae: Pholcidae). Zool Meded 79:61–114

Huber BA (2011a) Phylogeny and classification of Pholcidae (Araneae): an update. J Arachnol 39:211–222

Huber BA (2011b) Revision and cladistic analysis of *Pholcus* and closely related taxa (Araneae, Pholcidae). Bonner Zool Monogr 58:1–509

Huber BA (2012) Revision and cladistic analysis of the Afrotropical endemic genus *Smeringopus* Simon, 1890 (Araneae: Pholcidae). Zootaxa 3461:1–138

Huber BA (2022) Revisions of *Holocnemus* and *Crossopriza*: the spotted-leg clade of Smeringopinae (Araneae, Pholcidae). Eur J Taxon 795:1–241

Kovács G, Szinetár C, Eichardt J (2008) Data on the biology of pale cellar spider (*Spermophora senoculata* [Dugés, 1836]) (Araneae: Pholcidae). A Nyme Savaria Egyetemi Központ Tudományos Közleményei XVI. Természettudományok 11(Biológia):125–135

Nentwig W (1983) The prey of web-building spiders compared with feeding experiments (Araneae: Araneidae, Linyphiidae, Pholcidae, Agelenidae). Oecologia 56:132–139

Sedey KA, Jakob EM (1998) A description of an unusual dome web occupied by egg-carrying *Holocnemus pluchei* (Araneae, Pholcidae). J Arachnol 26:385–388

Slowik J (2009) A review of the cellar spider genus *Psilochorus* Simon 1893 in America north of Mexico (Araneae: Pholcidae). Zootaxa 2144:1–53

Uhlenhaut H (2001) Beobachtungen zum Beutespektrum von Zitterspinnen (Pholcidae). Arachnol Mitt 22:37–41

Chapter 19: Salticidae

Bear A, Hasson O (1997) The predatory response of a stalking spider, *Plexippus paykulli*, to camouflage and prey type. Anim Behav 54:993–998

Freed AN (1984) Foraging behavior in the jumping spider *Phidippus audax*: bases for selectivity. J Zool 203:49–61

Guseinov EV (2004) Natural prey of the jumping spider *Menemerus semilimbatus* (Hahn, 1827) (Araneae: Salticidae), with notes on its unusual predatory behaviour. Arthropoda Selecta Spec Issue 1:93–100

Harm M (1969) Zur Spinnenfauna Deutschlands, VI. Revision der Gattung *Salticus* Latreille (Arachnida: Araneae: Salticidae). Senck Biol 50:205–218

Harm M (1981) Revision der mitteleuropäischen Arten der Gattung *Marpissa* C. L. Koch 1846 (Arachnida: Araneae: Salticidae). Senckenberg Biol 61:277–291

Jackson RR, MacNab A (1989) Display and predatory behaviour of *Plexippus paykulli*, a jumping spider (Araneae, Salticidae) from Florida. NZ J Zool 16:151–168

Logunov DV (1998) *Pseudeuophrys* is a valid genus of the jumping spiders (Araneae, Salticidae). Rev Arachnol 12:109–128

Ramseyer LJ, Crawford RL (2017) First records of the European species *Pseudeuophrys lanigera* (Simon, 1871) (Araneae: Salticidae) in North America. Pan-Pac Entomol 93:226–228

Taylor BB, Peck WB (1975) A comparison of northern and southern forms of *Phidippus audax* (Hentz) (Araneida, Salticidae). J Arachnol 2:89–99

Wesołowska W (1999) A revision of the spider genus *Menemerus* in Africa (Araneae: Salticidae). Genus 10:251–353

Chapter 20: Scytodidae

Bürgis H (1990) Die Speispinne *Scytodes thoracica* (Araneae: Sicariidae). Ein Beitrag zur Morphologie und Biologie. Mitt Pollichia 77:289–313

Clements R, Li D (2005) Regulation and non-toxicity of the spit from the pale spitting spider *Scytodes pallida* (Araneae: Scytodidae). Ethology 111:311–321

Dabelow S (1958) Zur Biologie der Leimschleuderspinne *Scytodes thoracica* (Latreille). Zool Jahrb (Syst) 86:85–126

Nentwig W (1985b) Feeding ecology of the tropical spitting spider *Scytodes longipes* (Araneae, Scytodidae). Oecologia 65:284–288

Suter R, Stratton G (2013) Predation by spitting spiders: elaborate venom gland, intricate delivery system. In: Nentwig W (ed) Spider ecophysiology. Springer, Berlin

Chapter 21: Segestriidae

Brignoli PM (1976) Spiders of Italy XXIV. Notes on the morphology of internal genitalia of Segestriidae and hints on the Italian species. Fragm Entomol 12:19–62

Giroti AM, Brescovit AD (2011) The spider genus *Segestria* Latreille, 1804 in South America (Araneae: Segestriidae). Zootaxa 3046:59–66

Schmidt G (1990) Zur Spinnenfauna der Kanaren, Madeiras und der Azoren. Stuttgarter Beitr Naturkd (A) 451:1–46

Chapter 22: Selenopidae

Crews SC, Wienskoski E, Gillespie RG (2008) Life history of the spider *Selenops occultus* Mello-Leitão (Araneae, Selenopidae) from Brazil with notes on the natural history of the genus. J Nat Hist 42–43:2747–2761

Kunt KB, Tezcan S, Yağmur EA (2011) The first record of family Selenopidae (Arachnida: Araneae) from Turkey. Turk J Zool 35:607–610

Leroy A, Leroy J (2012) Spiders of southern Africa. Struik Publishers, Cape Town, 96 pp

Zamani A, Crews SC (2019) The flattie spider family Selenopidae (Araneae) in the Middle East. Zool Middle East 65:79–87

Chapter 23: Sicariidae

Gertsch WJ, Ennik F (1983) The spider genus *Loxosceles* in North America, Central America, and the West Indies (Araneae, Loxoscelidae). Bull Am Mus Nat Hist 175:264–360

Nentwig W, Kuhn-Nentwig L (2013a) Spider venoms potentially lethal to humans. In: Nentwig E (ed) Spider ecophysiology. Springer, Berlin

Nentwig W, Pantini P, Vetter RS (2017) Distribution and medical aspects of *Loxosceles rufescens*, one of the most invasive spiders of the world (Araneae: Sicariidae). Toxicon 132:19–28

Vetter RS (2008) Spiders of the genus *Loxosceles* (Araneae, Sicariidae): a review of biological, medical and psychological aspects regarding envenomations. J Arachnol 36:150–163

Vetter RS (2015) The brown recluse spider. Comstock Publishing Associates, Ithaca

White J, Cardoso JL, Fan HW (1995) Clinical toxicology of spider bites. In: Meier J, White J (eds) Clinical toxicology of animal venom and poisons. CRC, Boca Raton

Chapter 24: Sparassidae

Isbister GK, Hirst D (2003) A prospective study of definite bites by spiders of the family Sparassidae (huntsmen spiders) with identification to species level. Toxicon 42:163–171

Jäger P (2000) Selten nachgewiesene Spinnenarten aus Deutschland (Arachnida: Araneae). Arachnol Mitt 19:49–57

Jäger P (2014) *Heteropoda* Latreille, 1804: new species, synonymies, transfers and records (Araneae: Sparassidae: Heteropodinae). Arthropoda Selecta 23:145–188

Ross J, Richman DB, Mansour F, Trambarulo A, Whitcomb WH (1982) The life cycle of *Heteropoda venatoria* (Linnaeus) (Araneae: Heteropodidae). Psyche Camb 89:297–305

Chapter 25: Theridiidae

Ayoub NA, Friend K, Clarke T, Baker R, Correa-Garhwal SM, Crean A, Dendev E, Foster D, Hoff L, Kelly SD, Patterson W, Hayashi CY, Opell BD (2021) Protein composition and associated material properties of cobweb spiders' gumfoot glue droplets. Integr Comp Biol 61:1459–1480

Fox EGP, Brescovit AD, Solis DR, Jesus CM, Bueno O (2009) Immatures of *Nesticodes rufipes* (Araneae, Theridiidae) causing considerable damage to ant colonies in the laboratory. Sociobiology 53:75–76

Guido G (2010) *Nesticodes rufipes* – Erstnachweis einer pantropischen Kugelspinne in Deutschland (Araneae: Theridiidae). Arachnol Mitt 39:40

Isbister GK, Gray MR (2003) Latrodectism: a prospective cohort study of bites by formally identified redback spiders. Med J Aus 179:88–91

Kaston BJ (1970) Comparative biology of American black widow spiders. Trans San Diego Soc Natl Hist 16:33–82

Knoflach B, Pfaller K (2004) Cobweb spiders – an introduction (Araneae, Theridiidae). In: Thaler K (ed) Diversity and biology of web spiders, scorpions and other arachnids. Denisia 12:111–160

Levi HW (1957a) The spider genera *Crustulina* and *Steatoda* in North America, Central America, and the West Indies (Araneae, Theridiidae). Bull Mus Comp Zool 117:367–424

Levi HW (1957b) The spider genera *Enoplognatha*, *Theridion*, and *Paidisca* in America north of Mexico (Araneae, Theridiidae). Bull Am Mus Nat Hist 112:1–124

Levi HW (1962) The spider genera *Steatoda* and *Enoplognatha* in America (Araneae, Theridiidae). Psyche Camb 69:11–36

Levi HW (1967) Cosmopolitan and pantropical species of theridiid spiders (Araneae: Theridiidae). Pac Insects 9:175–186

Levy G (1998) Araneae: Theridiidae. In: Fauna Palaestina, Arachnida III. Israel Academy of Sciences and Humanities, Jerusalem, 228pp

Levy G, Amitai P (1982) The comb-footed spider genera *Theridion*, *Achaearanea* and *Anelosimus* of Israel (Araneae: Theridiidae). J Zool Lond 196:81–131

Levy G, Amitai P (1983) Revision of the widow-spider genus *Latrodectus* (Araneae: Theridiidae) in Israel. Zool J Linnean Soc 77:39–63

Lotz LN (1994) Revision of the genus *Latrodectus* (Araneae: Theridiidae) in Africa. Navorsinge Nas Mus Bloemfontein 10:1–60

Nentwig W, Kuhn-Nentwig L (2013b) Spider venoms potentially lethal to humans. In: Nentwig W (ed) Spider ecophysiology. Springer, Berlin

Paterson Fox EG, Brescovit AD, Solis DR, de Jesus CM, Bueno OC (2009) Immatures of *Nesticodes rufipes* (Araneae, Theridiidae) causing considerable damage to ant colonies in the laboratory. Sociobiology 53:71–77

Rossi MN, Godoy WAC (2006) Prey choice by *Nesticodes rufipes* (Araneae, Theridiidae) on *Musca domestica* (Diptera, Muscidae) and *Dermestes ater* (Coleoptera, Dermestidae). J Arachnol 34:186–193

Sekhar R, Sunil JK (2017) New distributional records of *Nesticodes rufipes* (Lucas, 1846) in India. Munis Entomol Zool 12:478–480

Wiehle H (1937b) Spinnentiere oder Arachnoidea. 26. Familie. Theridiidae oder Haubennetzspinnen (Kugelspinnen). Die Tierwelt Deutschl 33:119–222

Chapter 26: Uloboridae

Brignoli PM (1979) Contribution to the knowledge of Palaearctic Uloboridae (Araneae). Rev Arachnol 2:275–282

Hänggi A, Straub S (2016) Storage buildings and greenhouses as stepping stones for non-native potentially invasive spiders (Araneae) – a baseline study in Basel, Switzerland. Arachnol Mitt 51:1–8

Klein W, Stock M, Wunderlich J (1994) Zwei nach Deutschland eingeschleppte Spinnenarten (Araneae) – *Uloborus plumipes* Lucas und *Eperigone eschatologica* (Bishop) – Als Gegenspieler der weißen Fliege im geschützten Zierpflanzenbau? Beitr Araneol 4:301–306

Opell BD (1979) Revision of the genera and tropical American species of the spider family Uloboridae. Bull Mus Comp Zool 148:443–549

Chapter 27: Zoropsidae

Eggs B, Wolff JO, Kuhn-Nentwig L, Gorb SN, Nentwig W (2015) Hunting without a web: how lycosoid spiders subdue their prey. Ethology 121:1166–1177

Griswold CE, Ubick D (2001) Zoropsidae: a spider family newly introduced to the USA (Araneae, Entelegynae, Lycosoidea). J Arachnol 29:111–113

Hänggi A, Zürcher I (2013) *Zoropsis spinimana* — a Mediterranean spider has become native in Basel (NW-Switzerland). Mitt Naturf Ges Beider Basel 14:125–134

Nentwig W, Gnädinger M, Fuchs J, Ceschi A (2013b) A two year study of verified spider bites in Switzerland and a review of the European spider bite literature. Toxicon 73:104–110

Thaler K, van Harten A, Knoflach B (2006) *Zoropsis saba* sp. n. from Yemen, with notes on other species (Araneae, Zoropsidae). Bull Br Arachnol Soc 13:249–255

Index